ENVIRONMENTAL SCIENCE, ENGINEERING AND TECHNOLOGY

# DISPERSANTS IN OIL SPILLS

## ISSUES AND RESEARCH

# ENVIRONMENTAL SCIENCE, ENGINEERING AND TECHNOLOGY

Additional books in this series can be found on Nova's website under the Series tab.

Additional E-books in this series can be found on Nova's website

ENVIRONMENTAL SCIENCE, ENGINEERING AND TECHNOLOGY

# DISPERSANTS IN OIL SPILLS

## ISSUES AND RESEARCH

EDWARD K. HOLM
EDITOR

Nova Science Publishers, Inc.
*New York*

For permission to use material from this book please contact us:
Telephone 631-231-7269; Fax 631-231-8175
Web Site: http://www.novapublishers.com



**LIBRARY OF CONGRESS CATALOGING-IN-PUBLICATION DATA**

Dispersants in oil spills : issues and research / editor, Edward K. Holm.
p. cm. -- (Environmental science, engineering, and technology)
Includes bibliographical references and index.
ISBN 978-1-61209-241-6 (hardcover : alk. paper)
1. Oil spills--Cleanup--Mexico, Gulf of. 2. Dispersing agents--Toxicity testing--Mexico, Gulf of. 3. Dispersing agents--Environmental aspects--Mexico, Gulf of. I. Holm, Edward K.
TD427.P4D5465 2011
628.1'6833--dc22

2010047003

*Published by Nova Science Publishers, Inc. + New York*

# CONTENTS

# PREFACE

This new book examines the issues raised by the use of dispersants in the Deepwater Horizon spill. Dispersants change the distribution, not the amount, of oil within a marine environment. The author of this book examines how well the government handled the dispersant issues it faced in the absence of necessary scientific information and pursuant to a regulatory regime that had failed to anticipate this kind of problem. Also discussed is how, in light of lessons learned from this recent experience, government procedures and existing laws might be improved to allow for sounder decisions regarding the use of dispersants in the future.

Chapter 1- This Working Paper examines the issues raised by the use of dispersants in the Deepwater Horizon spill. Dispersants change the distribution, not the amount, of oil within a marine environment. They are chemicals typically applied directly to oil on the water surface in order to break the oil into small droplets that can then mix with water below the surface. The dispersed oil is rapidly diluted, mixing both vertically and horizontally in the water column.[1] While this alleviates high concentrations at the surface, it may expose organisms to lower, but more widespread, concentrations of oil.

Chapter 2- The U.S. Environmental Protection Agency's Office of Research and Development was asked to evaluate the cytotoxicity and potential for interaction with the androgen and estrogen receptors (AR, ER) of eight oil spill dispersants being used, or could be considered for use, in the Gulf of Mexico. These are Corexit 9500 (the current product being used), DISPERSIT SPC 1000, JD 2000, Nokomis 3-F4, Nokomis 3-AA, SAF-RON GOLD, Sea Brat #4, and ZI-400. To address this request, ORD staff and outside collaborators carried out a number of separate studies that were run using *in vitro* (cell-based) assays. A total of 8 cytotoxicity assays, 3 AR agonist assays, 1 AR antagonist assay and 4 ER agonist assays were run on the 8 dispersants, plus reference compounds. Tests were run across a wide range of dispersant concentrations (0.001 to 10,000 parts per million, or ppm). Two dispersants showed a weak signal in one of the four ER assays, but integrating over all of the ER and AR results these data do not indicate that any of the eight dispersants display biologically significant endocrine activity via the androgen or estrogen signaling pathways. All of the dispersants showed cytotoxicity in at least one cell type at concentrations between 10 and 1000 ppm. Both JD 2000 and SAF-RON GOLD tend to be less cytotoxic than the other dispersants. Likewise, DISPERSIT SPC 1000 tends to be more cytotoxic than the other dispersants in the cell-based assays.

Chapter 3- An estimated 4.9 million barrels of South Louisiana sweet crude (LSC) oil was released into the northern Gulf of Mexico between April 20 and July 15, 2010 caused by the explosion and collapse of the Deepwater Horizon oil exploration platform. The use of dispersants in oil spill response involves tradeoffs between effects to the shoreline and effects to pelagic and deep sea environments. In an effort to mitigate the impact of floating oil on sensitive shoreline habitats the decision was made to use dispersants listed on the U.S. Environmental Protection Agency's (EPA) National Contingency Plan (NCP) Product Schedule (EPA 2010a). Dispersants were applied offshore on the surface as well as underwater at the source of the leak. In an effort to assess options to address this unprecedented event, the EPA conducted independent studies with eight dispersants on the NCP Product Schedule to assess the relative acute toxicity of each dispersant alone, LSC oil alone and dispersant-LSC oil mixtures.

Chapter 4- Large quantities of Louisiana sweet crude oil have been released into the Gulf of Mexico since the explosion of the Deepwater Horizon oil exploration platform on April 20, 2010. As part of the integrated response effort to mitigate the impact of the oil in the environment, the decision was made to use dispersants listed on the U.S. Environmental Protection Agency's (EPA) National Contingency Plan (NCP) Product Schedule (EPA 2010a). Dispersants are being applied offshore on the surface as well as underwater at the source of the leak. The EPA conducted independent studies to assess the relative acute toxicity of eight dispersants on the NCP Product Schedule.

Chapter 5- EPA conducted acute toxicity tests of 1).multiple concentrations of Louisiana Sweet Crude Oil alone, and 2.) combinations of Louisiana Sweet Crude Oil with each of the eight dispersants. The organisms tested were two Gulf of Mexico aquatic species: (1) the mysid shrimp,

*Americamysis bahia*, an aquatic invertebrate, and (2) the inland silverside, *Men idia beryllina*, a small estuarine fish. These species are standard test organisms used in toxicity test for a variety of pollutants. The tests were conducted on mixtures of Louisiana Sweet Crude Oil and eight dispersant products found on the National Contingency Plan Product Schedule – Dispersit SPC 1000, Nokomis 3-F4, Nokomis 3-AA, ZI-400, SAFRON Gold, Sea Brat #4, Corexit 9500 A and JD 2000.

Chapter 6- The U.S. Environmental Protection Agency (EPA) is charged by Congress with protecting the Nation's land, air, and water resources. Under a mandate of national environmental laws, the Agency strives to formulate and implement actions leading to a compatible balance between human activities and the ability of natural systems to support and nurture life. To meet this mandate, EPA's research program is providing data and technical support for solving environmental problems today and building a science knowledge base necessary to manage our ecological resources wisely, understand how pollutants affect our health, and prevent or reduce environmental risks in the future.

In: Dispersants in Oil Spills: Issues and Research ISBN: 978-1-61209-241-6
Editors: Edward K. Holm 

*Chapter 1*

# THE USE OF SURFACE AND SUBSEA DISPERSANTS DURING THE BP DEEPWATER HORIZON OIL SPILL

***National Commission on the BP Deepwater Horizon Oil Spill and Offshore Drilling***

**Staff Working Paper No. 4**

*Staff Working Papers are written by the staff of the National Commission on the BP Deepwater Horizon Oil Spill and Offshore Drilling for the use of members of the Commission. They are preliminary, subject to change, and do not necessarily reflect the views either of the Commission as a whole or of any of its members. In addition, they may be based in part on confidential interviews with government and non-government personnel.*

This Working Paper examines the issues raised by the use of dispersants in the Deepwater Horizon spill. Dispersants change the distribution, not the amount, of oil within a marine environment. They are chemicals typically applied directly to oil on the water surface in order to break the oil into small droplets that can then mix with water below the surface. The dispersed oil is rapidly diluted, mixing both vertically and horizontally in the water column.[1] While this alleviates high concentrations at the surface, it may expose organisms to lower, but more widespread, concentrations of oil.

The use of dispersants in the aftermath of the Macondo deepwater well explosion was controversial for three reasons.[2] First, the total amount of dispersants used was unprecedented: 1.84 million gallons. Second, 771,000 of those gallons were applied at the wellhead, located 5,067 feet below the surface. Little or no prior testing had been done on the effectiveness and potential adverse environmental consequences of subsea dispersant use, let alone at those volumes.[3] Third, the existing federal regulatory system pre-authorized dispersant use in the Gulf of Mexico without any limits or guidelines as to amounts or duration. Faced with an emergency, the government had to make decisions about high-volume and subsea dispersant use within time frames that denied officials the opportunity to gather necessary information. The resulting uncertainty even fueled unfounded suspicions that BP was using dispersants without authorization from the government in an effort to mask the oil and to limit its ultimate liability.

This paper considers two issues. The first is how well the government handled the dispersant issues it faced in the absence of necessary scientific information and pursuant to a regulatory regime that had failed to anticipate this kind of problem. The second is how, in light of lessons learned from this recent experience, government procedures and existing laws might be improved to allow for sounder decisions regarding the use of dispersants in the future.

The paper is divided into three parts. Part I provides background information on dispersants and their potential authorization for use in responses to oil spills. It then recaps the chronology of the use of dispersants following the M acondo well explosion. This chronology includes the volume of dispersants used, where they were used, the types of dispersants used, and the role of various government agencies in making relevant decisions regarding dispersant use. Part I also describes some of the contemporaneous public controversy concerning the use of dispersants, including the debate, still ongoing, regarding their potentially adverse impacts. Part II considers the distinct questions of whether the government's decisions were reasonable at the time; and whether, regardless of their reasonableness or unreasonableness when made, preliminary scientific research since undertaken suggests those decisions may, in fact, have been sensible. Finally, Part III describes some possible implications for changes in agency procedures and regulations arising out of the use of dispersants in the Deepwater Horizon spill response that Commissioners may wish to consider.

# I. Background on Dispersants

## A. The Trade-offs of Dispersant Use

When an oil spill occurs, responders have several tools to manage potential environmental impacts. Mechanical means are generally preferred, but they cannot always be used and do not recover all of the spilled oil.[4] Non-mechanical methods such as in-situ burning and chemical dispersants can contribute to the elimination of the oil. In response to the Deepwater Horizon oil spill, BP used large amounts of the dispersant Corexit 9500 and some Corexit 9527.[5]

Dispersants function like detergents to break up oil into small droplets that mix easily with water. They contain a combination of surfactants and solvents. Surfactants are compounds that have lipophilic groups, which mix with non-polar substances like oil, and hydrophilic groups, which mix with polar substances like water. By combining lipophilic and hydrophilic groups, surfactants can lower surface tension to allow water and oil to mix more easily.[6] The solvents help the surfactants pass through the oil to reach the oil-water boundary where the surfactants operate.[7]

The resulting oil/water mixture takes the form of small droplets of dispersant-covered oil, which, because of their small size, can remain suspended in water rather than rising to the surface.[8] These droplets can move into and through the water column from the water's surface.[9] This process depends on outside forces to disperse the oil droplets through the top of the water column. For that reason, dispersants applied to surface oil slicks are more effective in areas with high wave energy.[10]

The toxicity of available dispersants has diminished substantially over the past several decades.[11] Generally, dispersants are less toxic than oil or chemically dispersed oil. However, dispersants and dispersed oil are typically more toxic than oil alone to embryos and larvae.[12]

Using dispersants to remove oil from the water surface has several potential benefits. First, less oil will float ashore to adversely affect shorelines and fragile estuarine environments. Second, animals and birds that float on or wade through the water surface may be less exposed to oil.[13] Third, dispersants may accelerate the rate at which oil biodegrades. Smaller droplets have a larger surface-area-to-volume ratio, which in theory should allow microorganisms greater access to the oil, and speed their rate of consumption.

The expected acceleration of this biodegradation is often cited as a major reason to use dispersants.

There are uncertainties regarding both the actual realization of some of these benefits, especially in the subsea, and potential offsetting costs. For instance, less oil on the surface means more in the water column, increasing exposure for subsurface marine life. And, while the smaller droplets may accelerate biodegradation, their smaller size increases the dissolution of potentially toxic compounds and exposure to aquatic organisms. Moreover, according to at least some scientific literature, the assumption of increased biodegradation may not always be accurate. Some studies have found that dispersants have no effect on the biodegradation rate or may even inhibit biodegradation.[14] It is also only largely in the aftermath of the Macondo well explosion that scientists have begun to research the extent to which oil-eating bacteria are present at the low temperatures of deepwater.[15] Finally, there is no reason to suppose that all dispersants act in the same manner. They may, depending upon their chemical makeup, have strikingly dissimilar impacts. For example, some evidence indicates that the ionic surfactant in Corexit 9527 and 9500 inhibits biodegradation while their non-ionic surfactants increase biodegradation.[16]

## B. Regulation of the Use of Dispersants in Oil Spill Response

The Clean Water Act expressly contemplates the use of dispersants in response to oil spills. Section 311(d)(2)(G) of the Act requires that the federal National Contingency Plan for oil spill response contain a schedule identifying:

> (i) dispersants . . . , if any, that may be used in carrying out the Plan,
> (ii) the waters in which such dispersants . . . may be used, and
> (iii) the quantities of such dispersant . . . which can be used safely in such waters . . . .[17]

In addition, subsection (G) requires each schedule to provide for use of other, non-listed dispersants: "[T]he President, or his delegate, may, on a case-by-case basis, identify the dispersants, other chemicals, and other spill mitigating devices and substances which may be used, the waters in which they may be used, and the quantities which can be used safely in such waters."[18]

The National Contingency Plan under the Clean Water Act and Oil Pollution Act of 1990 further provides for the establishment of regional and area-wide contingency plans, which may expressly pre-authorize the use of dispersants:

> In meeting the provisions of this paragraph, preauthori zati on plans may address factors such as the potential sources and types of oil that might be spilled, the existence and location of environmentally sensitive resources that might be impacted by spilled oil, available product and storage locations, available equipment and adequately trained operators, and the available means to monitor product application and effectiveness . . . . Approved preauthorization plans shall be included in the appropriate RCPs and ACPs [regional and area contingency plans].[19]

When dispersants have *not* been pre-authorized in an oil spill response contingency plan, they can still be approved after a spill has occurred. Federal regulations require the Federal On- Scene Coordinator to obtain "concurrence" in this circumstance from EPA and applicable state authorities, but require only "consultation" with the Department of Commerce (through the National Oceanic and Atmospheric Administration [NOAA]) and the Department of the Interior:

> [T]he OSC [On-Scene Coordinator], with the concurrence of the EPA representative to the RRT [Regional Response Team] and, as appropriate, the concurrence of the RRT representatives from the states with jurisdiction over the navigable waters threatened by the release or discharge, and in consultation with the DOC [Department of Commerce, i.e., NOAA] and DOI [Department of the Interior] natural resource trustees, when practicable, may authorize the use of dispersants, surface washing agents, surface collecting agents, bioremediation agents, or miscellaneous oil spill control agents on the oil discharge, provided that the products are listed on the NCP [National Contingency Plan] Product Schedule.[20]

The effect of pre-approval, accordingly, is to eliminate the need for approvals and consultations during the response and to allow the Federal On-Scene Coordinator to act unilaterally.

The National Contingency Plan also establishes "Area Committees"[21] under the direction of a Federal On-Scene Coordinator[22] that are charged with "work[ing] with State and local officials to expedite decisions for the use of dispersants and other mitigating substances and devices."[23]

The decision whether to approve the use of dispersants can be difficult, whether it occurs through the pre-approval process in developing a contingency plan or, in the absence of pre- approval, once a spill has occurred. As described by the National Research Council of the National Academies of Sciences, "[g]iven the potential impacts that dispersed oil may have on water-column and seafloor biota and habitats, thoughtful analysis is required prior to the spill event so that decision makers understand the potential impacts with and without dispersant application."[24] The trade-offs are complex:

> Decisions to use dispersants . . . involve trade-offs between decreasing the risk to water surface and shoreline habitats while increasing the potential risk to organisms in the water column and on the seafloor. This trade-off reflects the complex interplay of many variables, including the type of oil spilled, the volume of the oil, sea state and weather, water depth, degree of turbulence (thus mixing and dilution of the oil), and relative abundance and life stages of resident organisms.[25]

Under the National Contingency Plan, EPA is responsible for obtaining dispersant toxicity data from industry before placing a dispersant on the product schedule, which then serves as the basis for listing particular dispersants for pre-approved use in oil spill response contingency plans. The accuracy and consistency of pre-listing testing by manufacturers has been questioned, with toxicologists suggesting that the results of industry testing vary more widely than they should.[26]

Moreover, the required pre-authorization testing is limited to acute toxicity studies (48 and 96 hours) on two species: a fish species and a mysid shrimp species, *Menidia beryllina* and *Mysidopsis bahia*, respectively. EPA commonly uses these species in laboratory tests, and they are useful in providing comparative acute toxicity information, but the tests are not designed as proxies for all possible adverse ecosystem impacts. The pre-testing of dispersants did not include other important matters such as environmental persistence, effectiveness with multiple varieties of oil and at multiple temperatures, byproducts, and endocrine effects.

## C. The Use of Dispersants in Response to the Deepwater Horizon Spill

The federal government's response to the oil spill began immediately after the Macondo well explosion on the night of April 20, 2010. In addition to the emergency search-and-rescue mission, efforts to address the released oil were soon underway. Pursuant to the National Contingency Plan, the Coast Guard Captain of the nearest port, Morgan City, Louisiana, served as the Federal On-Scene Coordinator, in charge of the government's response action, until a few days later when the District (Eight) Commander, Rear Admiral Mary Landry, took over the Coordinator role.

The oil spill response contingency plans applicable to the Gulf (Regions 4 and 6 within the National Response Plan framework) pre-authorized the use of a list of specific dispersants. Neither of the plans limited the overall volume or duration of such pre-authorized use.[27] With the permission of the Federal On-Scene Coordinator, BP applied 14,654 gallons of the dispersant Corexit, which was on the approved list, on the surface during the week of April 20-26, 2010.[28] On April 29, 2010, the Coast Guard formally designated the Gulf spill a "Spill of National Significance" pursuant to the Oil Pollution Act.[29] Based on that designation, the Commandant of the Coast Guard, Admiral Thad Allen, became the "National Incident Commander" in charge of the federal government's response actions.

During May, dispersants were applied both to the surface and subsea, and the volume used increased rapidly. During the week of April 27 to May 3, 2010, responders applied 141,358 gallons to the surface, and that amount grew to 168,988 by the following week.[30] The week of May 11 to May 17, 2010, the amount of surface dispersants used reached 255,000 gallons.[31]

On May 1, 2010, Admiral Thad Allen reported that response crews had begun testing the subsurface application of dispersants to oil escaping from the broker riser.[32] Nearly 3,000 gallons of subsea dispersants were applied.[33] At the time, it was unclear whether the National Contingency Plan's pre-approval of the use of dispersants in the Gulf applied to subsea use in addition to surface use, and therefore whether additional EPA approval and NOAA consultation were required.[34] Notwithstanding those uncertainties regarding governing law, on May 7, 2010, "having deployed test applications of subsea dispersants, EPA halted subsea dispersant operations, awaiting additional test results."[35]

Testing and monitoring, however, presented substantial logistical and organizational problems. BP itself performed three tests, based on protocols established by EPA and the Coast Guard.[36] On May 15, 2010, the testing for effectiveness and toxicity that had been completed prompted EPA and the Coast Guard to announce their joint approval of subsea dispersant use with the condition that BP conduct further monitoring.[37] The EPA Administrator, Lisa Jackson, made the approval decision on behalf of EPA herself and has since publicly acknowledged the difficulty of making this decision with the limited amount of scientific information then available. Considerations related to response worker health and ease of application—subsea application would minimize the necessary human contact with dispersants, and could occur at night and in foul weather—reportedly played a role in the decision to approve the method.[38] By May 17, 2010, BP had made extensive use of dispersants. The cumulative totals by this time were 580,000 gallons on the surface and 45,000 gallons subsea.

On May 20, 2010, in the wake of continuing media reports relating public concern about the potential toxicity of the high volumes of dispersants being used,[39] the Coast Guard and EPA issued a joint directive requiring BP to identify and use a less toxic and more effective dispersant than Corexit 9500 from the list of dispersants authorized by the National Contingency Plan.[40] According to the data in the National Contingency Plan Product Schedule, some of the pre- approved dispersants were both less toxic and more effective on South Louisiana crude oil than Corexit 9500.[41] Based on the Plan, the federal directive required BP to identify a less toxic alternative to be used both on the surface and subsea at the source of the oil leak within 24 hours, and to begin using the less toxic dispersant within 72 hours of submitting the alternative. Specifically, the directive called for toxicity levels LD50 "at or below" 23 parts per million (ppm) for Menidia or 18 ppm for Mysidopsis.[42] If BP was unable to identify acceptable alternative dispersant products, BP had to provide the Coast Guard and EPA with a detailed description of the alternative dispersants investigated, and the reasons it believed those products did not meet the required standards.[43]

BP responded to the directive the day it was issued.[44] BP contended that only five products on the National Contingency Plan Product Schedule (which lists acceptable dispersants) met the criteria in the directive and that Corexit 9500A was the "best alternative."[45] BP noted that one of these five acceptable dispersants "contains a small amount of a chemical that may degrade to a nonyl phenol ," a class of chemicals that have been identified as potential endocrine disruptors and may persist in the environment for a period of years.[46] Unfortunately, BP said, neither the manufacturer nor BP had had the opportunity to test the product for these potential effects.[47]

BP said that it would be prudent to obtain the chemical formulas for the other dispersants to evaluate their potential to degrade to a nonylphenol, but indicated that it had not been able to do so.[48] BP noted that "there may be only limited information on the constituents of the dispersants, since the dispersants typically contain proprietary substances whose identities are not publicly available."[49] In contrast, BP explained, the manufacturer of Corexit had said that it reached "maximum biodegradability" within one month and was not persistent in the environment.[50] In short, BP concluded, Corexit "appears to have fewer long term effects than the other dispersants evaluated."[51] BP also made clear that the company did not, in any event, then have a sufficient stockpile of any dispersants other than Corexit and Sea Brat #4, and that the Sea Brat #4 supply might not be sufficient for both surface and subsea use.[52] Corexit 9500 was the only dispersant used during the remainder of the spill.

In a May 24, 2010 press conference, EPA Administrator Jackson stressed three points, while generally acknowledging that federal regulators remained "deeply concerned about the things we don't know" such as the "long-term effect on aquatic life."[53] First, she said, the government was instructing BP to "take immediate steps to significantly scale back the overall use of dispersants" and expressed EPA's belief that "we can reduce the amount of dispersant applied by as much as half, and I think probably 75%, maybe more."[54] Second, she expressed dissatisfaction with BP's efforts to analyze other dispersant options.[55] Third, she announced, EPA would perform its own tests to verify BP's data and to "determine the least toxic, most effective dispersant available in the volumes necessary for a crisis of this magnitude."[56]

Two days later, Administrator Jackson sent a letter to David Rainey of BP criticizing BP's inadequate compliance with the May 20, 2010 directive, which had instructed BP "to analyze alternative dispersants for toxicity and effectiveness and report back within 24 hours."[57] "Because we believe your analysis of potential alternative dispersants was insufficient," she wrote, "the EPA is performing its own scientific verification of the data BP presented."[58] EPA said it would make laboratory comparisons with Gulf of Mexico species, including a silverside fish and a mysid shrimp.[59] EPA would also identify a test for endocrine disrupters.[60] Administrator Jackson's letter continued: "Furthermore, as we discussed, the federal government, led by the Coast Guard, is reiterating its instructions to BP to take immediate steps to significantly scale back the overall use of dispersants."[61] A May 26, 2010 directive provided that "BP shall eliminate the surface application of dispersants" except in "rare cases where there may have to be an exemption."[62]

On June 30, 2010, EPA released results of its own testing of eight dispersants.[63] EPA had conducted acute toxicity tests with two Gulf of Mexico aquatic species, and *in vitro* cytotoxicity (cell damage) and endocrine screening assays using human cell lines. EPA's results indicated that none of the eight dispersants displayed significant endocrine disrupting activity. It also suggested that Corexit 9500 was not overall more toxic than alternatives: "While the dispersant products alone—not mixed with oil—have roughly the same impact on aquatic life, JD-2000 and Corexit 9500 were generally less toxic to small fish and JD-2000 and SAF-RON GOLD were least toxic to mysid shrimp."[64]

The effort to scale back use of dispersants had some effect. During the week of May 18, 2010, BP applied 190,000 gallons total .[65] The following week, it applied roughly two-thirds as much (135,000 gallons).[66] Surface use fell from 120,000 gallons the week of May 18, 2010, to 40,000 gallons the week of May 25, 2010, although it then rose again and remained steady for several weeks at 80-90,000 gallons per week. By the end of May, BP had used a total of 950,000 gallons of dispersants, of which 740,000 were applied on the surface and 210,000 subsea.[67] As the following table shows, use of dispersants remained at a roughly constant level through most of June, but then began to decline again later in the month through early July:

**Table 1. Weekly Use of Dispersants June 1-July 12[68]**

| Week | Weekly Use of Dispersants (gallons) | Total Use of Dispersants to Date (gallons) (lower bound)[69] |
|---|---|---|
| June 1 – June 7 | 171,000 (total):<br>50,000 (surface)<br>121,000 (subsea) | 1.12 million (total):<br>790,000 (surface)<br>331,000 (subsea) |
| June 8 – June 14 | 163,000 (total):<br>92,000 (surface)<br>71,000 (subsea) | 1.28 million (total):<br>882,000 (surface)<br>402,000 (subsea) |
| June 15 – June 21 | 169,000 (total):<br>88,000 (surface)<br>80,000 (subsea) | 1.45 million (total):<br>970,000 (surface)<br>482,000 (subsea) |
| June 22 – June 28 | 112,000 (total):<br>30,000 (surface)<br>83,000 (subsea) | 1.56 million (total):<br>1 million (surface)<br>565,000 (subsea) |
| June 29 – July 5 | 145,000 (total):<br>40,000 (surface)<br>92,000 (subsea) | 1.71 million (total):<br>1.06 million (surface)<br>645,000 (subsea) |
| July 6 – July 12 | 90,000 (total)<br>10,000 (surface)<br>90,000 (subsea) | 1.8 million (total)<br>1.07 (surface)<br>735,000 (subsea) |

Despite the joint Coast Guard-EPA directive that BP "eliminate the surface application of dispersants" except in "rare cases where there may have to be an exemption," the use of surface dispersants was not eliminated after May 26, 2010. The "rare cases" were not very rare. Until late June, surface use in most weeks remained at about 40% of the pre-directive rate. The directive remained in effect despite suggestions that it be modified as responders became aware that the oil flow was much larger than previously believed.[70]

After the well was capped on July 15, 2010, there was virtually no further use of dispersants. By that time, BP had applied a total of 1.84 million gallons, of which 1.07 million gallons were applied on the surface and 771,000 gallons were subsea.[71] As Table 1 indicates, the amounts injected underwater became larger than the amounts applied to the surface during the last three weeks. The week before the well was capped, only about 10% of dispersants used were applied to the surface. The National Incident Command estimated in an August 4, 2010 report that approximately 8% of the oil emanating from the well was chemically dispersed, either subsea or on the surface, while about three times that much evaporated or dissolved, and twice that much was dispersed naturally at the wellhead.[72]

Although the use of dispersants had ended earlier in the month, debate about the use of dispersants surfaced again at the end of July. On July 12, 2010, Admiral Allen's Chief of Staff informed Rep. Edward Markey that dispersants were used "only when absolutely necessary to preserve the health and safety of workers at the well site and to minimize shoreline impacts."[73] On July 30, 2010, Rep. Markey sent a letter to Admiral Allen pointing to more than 74 BP exemption requests in 48 days, of which all but ten were fully approved by the Coast Guard. Rep. Markey alleged "these applications appear to be rubber stamped by the Coast Guard."[74]

The next day, in a conference call, Admiral Allen and Administrator Jackson replied that they had cooperated closely and nearly attained the goal of a 75% reduction in dispersant use.[75] On August 1, 2010, Admiral Allen said in a press conference that field commanders on a caseby-case basis decided to use dispersants where surveillance aircraft spotted oil and no other method of cleaning it up was available in the area. Admiral Allen noted that the decision to use the dispersants did not rest with BP. Rather, he said, "it's a decision by the Federal on-scene coordinator" through a "very disciplined doctrinal process."[76]

In a *CNN* interview the following day, Admiral Allen elaborated upon the working relationship between Coast Guard and EPA regarding the use of dispersants. According to the Admiral, he and Administrator Jackson "talk daily about dispersant use," the Coast Guard "ha[s]n't ignored EPA's guidelines," and he was "satisfied" with dispersant use in the Deepwater Horizon disaster.[77] Relatedly, CNN quoted an EPA spokesman as saying that, "[w]hi l e EPA may not have concurred with every individual waiver granted by the federal on-scene coordinator, the agency believes dispersant use has been an essential tool in mitigating this spill's impact, preventing millions of gallons of oil from doing even more damage to sensitive marshes, wetlands and beaches and the economy of the Gulf coast."[78] These statements suggested that coordination with EPA did not mean that the Federal On-Scene Coordinator heeded EPA's advice on all occasions. Given the pre-approval of dispersant use, the Federal On-Scene Coordinator was not required to do so.

## II. Assessing the Federal Government's Use of Dispersants during the Deepwater Horizon Spill

It is too early to assess many aspects of the federal government's use of dispersants in response to the Deepwater Horizon spill between the explosion on April 20, 2010, and the containment of the well on July 15, 2010. In making any assessment, moreover, it is important to distinguish between three inquiries: (1) whether the federal government adequately prepared in advance for the possible use of dispersants to address such a spill; (2) whether, once the spill occurred, the government's decisions regarding the use of dispersants were reasonable in light of the resources and the information then available; and (3) whether, with the benefit of hindsight, those government decisions, regardless of their reasonableness or unreasonableness when made, resulted in benefits that outweighed harms.

### A. The Adequacy of the Government's Contingency Planning

The first of these three questions is the one most easily answered. The government was not adequately prepared for the use of dispersants to address such a large oil spill. Notwithstanding the National Contingency Plan's express requirements for planning regarding the use of dispersants, including pre-authorization to deal with emergencies, EPA clearly did not anticipate the potential demands of an oil spill of the kind the nation faced after the Macondo well explosion. In particular, EPA did not consider, in its roles on the National

Response Team and the relevant Regional Response Teams, the possibility that dispersants might have to be used in the massive volumes required in the Gulf. And EPA did not consider the distinct possibility that massive volumes of dispersants might be needed at the subsea level.

Neither omission can be justified on the ground that a major subsea spill was wholly unforeseeable. The oil and gas industry has been extracting high volumes of oil from reservoirs in the Gulf for twenty years. This is not a new, unanticipated development. Nor is deepwater drilling. Yet, just as the Minerals Management Service and industry failed to plan adequately for a blowout of this magnitude and duration (topics to be discussed in other staff work), EPA did not consider adequately the challenges of dispersant use flowing from large-scale drilling operations, especially operations in deepwater. EPA did not require studies or testing that took account of the likely amounts or locations of dispersant use necessary in the event of a well blowout, in particular a deepwater blowout.

Nor had NOAA adequately planned for such an event. NOAA has significant responsibility to provide scientific support for national response and contingency planning pursuant to the Clean Water and Oil Pollution Acts.[79] Its related expertise arises out of its work in many areas, including its duties under the Endangered Species Act,[80] Marine Mammal Protection Act,[81] and Magnuson-Stevens Fishery Conservation and Management Act[82] to protect endangered and threatened species of marine life, marine mammals, and the nation's fisheries in U.S. waters. Yet NOAA had also not previously evaluated the potential impacts of voluminous and extended use of dispersants on marine life and the nation's fisheries.

As a result, the National Incident Commander, the EPA Administrator, and the NOAA Administrator were seriously handicapped when the Macondo well explosion occurred and decisions had to be made immediately in the absence of adequate contingency planning. These officials had to make difficult choices with insufficient information about the critical trade-offs identified by the National Academy of Sciences for the use of dispersants: the value of the dispersants in reducing the harm caused by released oil versus the potential risks of harm from the dispersants themselves. The limited toxicity data they possessed was questionable and limited to acute lethal effects on two estuarine species.[83] It did not consider potential environmental persistence resulting from repeated or continuous sublethal effects, such as endocrine disruption.

The absence of adequate contingency planning had a further negative impact on the effectiveness of the government's response. It made unclear the lines of authority between various federal agencies in determining whether dispersants should be used. In particular, there was uncertainty regarding the extent to which the Coast Guard needed to secure EPA's approval before permitting the use of dispersants. Notwithstanding the lack of any requirement that the Federal On-Scene Coordinator defer to EPA on the use of dispersants at the surface (given EPA's pre-approval), and the lack of a clear requirement with regard to the subsea, EPA decided to exercise substantial control over both types of dispersant use, which at times led to delays in necessary decision-making (and, according to Coast Guard responders, to some avoidable shoreline impacts from oil as a result of the inability to use dispersants quickly).[84]

## B. The Reasonableness of the Government's Decision to Authorize the Use of Dispersants at the Time It Was Made

As described above, the reasonableness of the federal government's decision to authorize the use of dispersants is distinct from the questions of whether there was adequate contingency planning (which there was not) and whether the decision ultimately turned out to be prudent. This inquiry instead focuses on whether the government acted reasonably given the limited knowledge and resources that it possessed at the time. Based on the information currently available to the Commission staff, we cannot conclude that the government acted unreasonably in deciding to approve the use of massive volumes of dispersants at the subsea and surface.

Because federal agencies had failed to plan adequately, they did not possess the scientific information that officials most certainly would have wanted to guide their choices. They had to make choices nevertheless: Millions of gallons of oil were flowing from the Macondo well into the Gulf of Mexico every day, imperiling the responders who worked in the immediate vicinity of the spill, residents living along the Gulf Coast, the Gulf marine ecosystem, and the fishing and tourism industries.

Given the conditions under which officials like Admiral Allen and EPA Administrator Jackson were acting, there is no clear evidence that their decisions to authorize high volumes of dispersants, including at the subsea, were unreasonable. They instead appear to have acted reasonably in the difficult circumstances in which they were placed. For instance, officials directed the Regional Response Teams to seek input as quickly as possible from fifty expert scientists.[85] On June 4, 2010, the experts reported a consensus that "use of dispersants and the effects of dispersing oil into the water column has generally been less environmentally harmful than allowing the oil to migrate on the surface into the sensitive wetlands and near shore coastal habitats."[86] In the experts' view, though gaps in relevant information existed, the environmental trade-off between the deep-ocean ecosystem and the shoreline made dispersants an acceptable choice.

There are, however, three caveats regarding the decision-making process. First, Commission staff heard repeated reports that EPA could have done a better job of ensuring that its on-scene representatives had both the expertise and the authority to make decisions regarding the use of dispersants, so as to avoid the delays that reportedly occurred because of the absence of such authority and expertise. For example, the Commission staff has heard from two sources that EPA waited until late June to permanently install one of the Agency's most senior officials at the Unified Command Center in Robert, Louisiana.[87]

The second caveat relates to implementation of the planned approach for decision- making regarding dispersants. The planning documents for the area require the Regional Response Teams to make decisions about novel uses of dispersants, upon request from the Federal On-Scene Coordinator. Here, as the issue of dispersant application became more and more prominent in the media and for the public, the decisions to apply both surface and subsea dispersants were taken out of hands of the Regional Response Teams. Admiral Allen and Administrator Jackson to a large extent bypassed the National and Regional Response Team structures and instead issued decisions regarding dispersant policy through joint directives. Though this reflected the high level at which the issues were being evaluated, it was outside of the process that responders were supposed to implement.

These two caveats aside, the Commission staff has reason to believe that there was generally a sound and cooperative working relationship between the federal agencies on the question of dispersants.[88] While the National and Regional Response Teams did not play the coordinating and decision-making role envisioned under the National Contingency Plan, the Federal On-Scene Coordinators worked directly with EPA and NOAA on dispersant policy. That coordination resulted in, among other things, the specific designation of subsea dispersants as an appropriate response technology subject to stringent limits on amounts as well as expansive testing and monitoring guidelines. In addition, the Federal On-Scene Coordinators and EPA worked together to reduce significantly the application of surface dispersants and to resolve the disagreements between the two agencies.

The third caveat relates to the role of BP. The fact that BP itself (or its oil spill response contractors) directly applied the dispersants authorized by the federal government led to the impression that BP rather than federal officials was in charge of decisions regarding dispersant use. Commission staff has not discovered any evidence that such a usurpation of government authority occurred. Nor could Commission staff conclude, based on interviews with Coast Guard responders, that BP or its contractors ever intentionally violated government directives regarding dispersant use (e.g., regarding the permitted locations for such use). Yet, the impression remained and fueled public distrust of the decision to use dispersants.

## C. The Prudence, in Hindsight, of the Federal Government's Decision to Permit Use of High Volumes of Dispersants at the Surface and Subsea

It is too soon to answer this final question with the degree of certainty necessary for scientific analysis. The gap between what federal government officials should know prior to the use of high volumes of dispersants at the surface and subsea and what they in fact know has begun to narrow. But closing that gap will require rigorous scientific inquiry based on years of data collection and analysis, followed by the essential process of peer review, before any conclusions can be drawn upon which future government officials can safely rely.

With this crucial limitation in mind, EPA's preliminary analyses do not suggest that the government's use of dispersants caused major problems. Just the opposite, they support the possibility that the benefits of dispersant use outweighed the costs.

First, it appears that the subsea use of dispersants served an important function by increasing the safety of the working conditions faced by responders in the immediate vicinity of the spill. That group included individuals working on containment efforts—to cap the well—and those seeking to retrieve, burn, and skim oil. The very real concern had been that high concentrations of volatile organic compounds within the oil would be a serious safety and health hazard to response workers. The use of subsea dispersants, by reducing the volume and concentration of the oil reaching the surface, likewise reduced those associated risks.[89]

Second, EPA's subsequent toxicity tests, while still preliminary, have not revealed major problems. On August 2, 2010, EPA released the results of additional tests on the toxicity of dispersants, which the Agency contended "confirm that the dispersant used in response to the oil spill in the gulf, Corexit 9500A, is no more or less toxic than the other available alternatives."[90] The EPA report itself concluded:

> Overall, the di spersants/L[ouisiana] S[weet] C[rude]mixtures were classified as being highly toxic to moderately toxic depending on the test species and dispersant. The ZI400/L[ouisiana] S[weet] C[rude] mixture was the exception and would be considered only slightly toxic to *Menidia*. Corexit 9500A, the dispersant that has been applied offshore at the surface and in the deep ocean, falls into the moderately toxic category for both species. For all eight dispersants in both test species, the dispersants alone were less toxic than the dispersant-oil mixture.[91]

Finally, EPA also reported on August 2, 2010, that the dispersants seemed to have succeeded in protecting the coastal area from greater contamination from the oil spill. The Agency referred to "fluorescence data that indicated the dispersants are working to keep the oil away from the shore. . . . [T]he dispersants are working to keep oil off our precious shorelines and away from sensitive ecosystems."[92] The Agency further noted that "EPA monitoring has not found dispersant chemicals in water or sediment near coasts or wetlands."[93]

On August 4, 2010, experts from NOAA, the National Institute of Standards and Technology, and the United States Geologic Survey released two reports that lent support to the claim that dispersants decreased the harms that might have otherwise resulted from the oil spill, by indicating that a significant percentage (8%) of the oil was chemically dispersed: the *Deepwater Horizon MC252 Gulf Incident Oil Budget* and a supporting document entitled *BP Deepwater Horizon Oil Budget: What Happened to the Oil* (collectively, the Oil Budget). 94 These reports have since been the subject of controversy for potentially overstating in significant respects and understating in other respects the amount of oil from the spill "remaining" in the Gulf.[95] One major focal point of criticism was the failure of the Oil Budget to analyze and take account of biodegradation, which chemical dispersants are intended in part to promote.[96] Several subsequent, non-governmental reports have debated how quickly subsurface Macondo oil is biodegrading.[97]

The ongoing debate regarding the fate of subsurface Macondo oil underscores the futility of trying now to evaluate conclusively the government's decision to use high volumes of dispersants. Even without application of dispersants subsea, the turbulent mixing of oil and gas from the wellhead could have created the deepwater plume of dispersed oil. According to the Oil Budget, subsea dispersant use only added to the amount of oil in the plume.[98] As more scientific research occurs, a better understanding of the degradation and impacts of this naturally and chemically dispersed oil will undoubtedly emerge.[99] It will also take time and research to determine whether the dispersants themselves, used in such high volumes and subsea, have any longer-term detrimental effects on marine life or public health. For now, there has not been compelling evidence of harmful effects to indicate that decision-makers misjudged the risks versus the benefits of applying dispersants.

# III. Issues for Commission Consideration

This final part describes policy implications for Commissioner consideration that arise from possible lessons learned from the use of dispersants during the Deepwater Horizon spill response. These lessons and related policy implications are not intended as exhaustive of those that may flow naturally from the above analysis, but merely illustrative of the possibilities.

## A. Further Research

Perhaps more than anything, the Deepwater Horizon experience with dispersants reveals the paucity of the kind of information that government officials need to make intelligent decisions about dispersant use in response to an oil spill. Although the absence of such information was well known before April 20, 2010, its practical effect had not been so glaringly realized.

As of 1999, EPA reported, "few long-term environmental effects tests have been conducted after a dispersant application."[100] In 2005, the National Research Council noted that U.S. research funding to support oil spill response was "extremely limited and declining" (with an annual total below $10 million).[101] The Council called on the relevant federal agencies to develop an integrated research plan focusing on peer-reviewed information.[102] Only a quarter of the $40 million in proposed research funding on dispersants and chemically dispersed oil ever materialized.[103]

The Deepwater Horizon oil spill confirms the urgency of these prior funding requests and suggests additional needs as well, including, for example, studies about the impacts of high volumes of dispersants, subsea impacts, and the long-term fate and effects of dispersants and dispersed oil—none of which appear to have been meaningfully addressed or at least addressed to the extent that now seems essential.[104] Efforts are ongoing to learn more about dispersant impacts in the Gulf. On August 3, 2010, the National Incident Commander recommended a detailed monitoring strategy, with NOAA as the operational lead, to evaluate the "distribution of indicators of break-down products of dispersants used in oil spill response activities."[105]

The development of dispersant alternatives should also be a priority. So-called "green chemistry" carries promise. Dispersants would seem to be a potentially important market for efforts to find new chemical products that are effective, less toxic, and more readily biodegradable.

Research and development, of course, requires funding. Offshore drilling provides a context within which substantial funding should be in reach. The nation's need for oil and gas from the outer continental shelf is undeniable. But so too are the massive revenues those reserves yield in the market and the harm, as recent events demonstrate, if drilling goes awry. The smallest fraction of those revenues, whether charged directly to industry or originating in what the government already receives, would provide a major benefit in terms of potential to mitigate the impact of oil spills from offshore drilling.

## B. Government Contingency Planning and Decision-Making Procedures

Government contingency planning for the use of dispersants was, as described, lacking. The federal agencies charged with planning did not adequately anticipate the need for dispersants in high volumes and at subsea locations. Federal officials must now survey existing and future offshore facilities and locations and consider systematically the particular challenges they present for spill response. In the future, officials should not find themselves similarly faced in the future with the need to make immediate decisions in the absence of adequate information.

Contingency planning reform should extend to rethinking both testing requirements and the use of pre-approved lists of dispersants. Plainly, the pre-approval process has significant advantages in the immediate aftermath of an oil spill and for that reason should not be abandoned. Indeed, it should be more rigorously applied by ensuring that those dispersants that are pre-approved are subject to more comprehensive testing.

There is clearly a need for expanded testing and greater information regarding dispersants placed on the National Contingency Plan Product Schedule, to include characteristics such as effectiveness and persistence under different environmental conditions. Testing should also be based on the use of higher volumes, including subsea. Moreover, current EPA protocols for industry testing may not be adequate to yield reliable and consistent results.[106] Given the ever- changing nature of the underlying science, periodic updating of testing and testing protocols is essential.

The Deepwater Horizon spill also suggests the possibility of including temporal, spatial, and/or volumetric limits on the pre-approval of dispersants for use in a geographic area. It is one thing to pre-approve based on the frequently reliable assumption that the response action will be limited in time, space, and dispersant volume. But, as the Deepwater Horizon spill dramatically illustrated, where those assumptions no longer hold, the force of a pre-approval is diminished. In particular, there is more reason to allow for federal officials other than the On-Scene Coordinator, such as EPA officials who possess particular expertise, to play a role in decision- making during the actual response. To that end, contingency planning for the use of dispersants during oil spill response should consider distinguishing between types of oil spills, based on their temporal duration, spatial reach, and volume.

With greater authority comes greater responsibility. During the Deepwater Horizon spill, there were reports that on-scene EPA representatives lacked the expertise and authority essential in a response action. Any enhancement of EPA's authority therefore must be coupled with assurances that EPA has the resources and clear lines of decision-making authority necessary for effective spill response. Ultimately, any recommendations for changes in the unified command structure should turn not just on the recent experience with the use of dispersants, but on a more cross-cutting inquiry, which is the subject of a separate Commission staff working paper. The issues surrounding dispersant use should inform that broader set of recommendations.

Finally, federal officials must from the outset leave no question in the public's mind regarding who is in charge during an emergency response, especially when, as happened with dispersants, public concern with the wisdom of the government's decisions is great. Here, a mistaken impression was created in the minds of too many that BP was making the decisions based on its own interests. That misimpression fueled the controversy over the potentially harmful impacts of dispersants, which itself harmed the public, creating real fears that had economic consequences to the extent they affected tourism and other consumer choices. In the future, government officials must leave no doubt that they, and not private industry, are making difficult decisions.

## End Notes

[1] Press Conference with Lisa P. Jackson, EPA Administrator (May 24, 2010), *available at* http://www.epa. gov/bpspi l l/di spersants/statement-di spersant-use-may24.pdf [hereinafter Jackson Press Conference]. For further information about dispersants, see Committee on Understanding Oil Spill Dispersants: Efficacy and Effects, National Research Council of the National Academies, UNDERSTANDING OIL SPILL DISPERSANTS: EFFICACY AND EFFECTS 9 (2005), *available at* http://www.nap.edu/catalog/11283.html [hereinafter NRC Report].

[2] The day after the Macondo well was capped and the amount of daily dispersant use dropped precipitously, a group of marine scientists opposed to the high volume use of dispersants issued a statement calling for an immediate end to their use. See Susan D. Shaw et al., *Consensus Statement: Scientists Oppose the Use of Dispersant Chemicals in the Gulf of Mexico* (July 16, 2010), available at http://www.meriresearch.org/Portals/0/Documents/ CONSENSUS%20STATEMENT%20ON%20DISPERSANTS% 20IN%20TH E%20GU L F%20updated%20July%2017.pdf.

[3] BP's "Lessons Learned" report refers to "limited trials" and "some discussion in technical papers of applying dispersant to the source." BP, DEEPWATER HORIZON CONTAINMENT AND RESPONSE: HARNESSING CAPABILITIES AND LESSONS LEARNED 26 (2010), *available at* http://www.boemre.gov/ooc/PDFs/NarrativeFinal.pdf. BP also claims that "EPA has permitted use of dispersants subsea to remediate oil spills since the 1990s." *Id.* None of the experts (whether from EPA, BP, or independent) with whom Commission staff spoke in researching this paper was aware of prior subsea use of dispersants, which suggests that any such use or approval was not well-known. On September 7, 2010, we requested information from BP's counsel regarding the trials, papers, and EPA actions to which BP refers. On September 27, 2010, BP's counsel confirmed in an email to Commission staff that "the reference on page 26 of the report to EPA's approval of subsea dispersant use in the 1990s is an error. Although dispersants have been on the Product Schedule since the 1970s, we understand that only surface uses were contemplated in the 1990s." BP's counsel also provided the Commission staff with references to two technical papers discussing the application of subsea dispersants. *See*, e.g., NRC Report at 138.

[4] *See* NRC Report at 9 ("The effectiveness of mechanical response techniques is variable and highly influenced by the size, nature, and location of the spill as well the environmental conditions under which the response is carried out. Essentially, mechanical response works satisfactorily under a finite subset of all possible spill scenarios.").

[5] *See* David Biello, Is Using Dispersants on the BP Gulf Oil Spill Fighting Pollution with Pollution?, SCI. AM. (June 18, 2010), *available at* http://www.scientificamerican.com/ article.cfm?id=is-using-dispersants-fighting-pollutionwith-pollution. Toxicity information on these dispersants can be found in Anita George-Ares & James R. Clark, *Aquatic Toxicity of Two Corexit Dispersants*, 40 CHEMOSPHERE 897 (2000).

[6] *See* Coastal Response Research Center, RESEARCH & DEVELOPMENT NEEDS FOR MAKING DECISIONS REGARDING DISPERSING OIL 1 (Apr. 2006), http://www.crrc.unh.edu/dwg/ dispersant_workshop_report-final.pdf [hereinafter CRRC Report].

[7] *See* International Tanker Owners Pollution Federation Ltd, Dispersants, *http://www.itopf.com/ spillresponse/cl* ean-up-and-response/di spersants/.

[8] *See* CRRC Report at 7.
[9] *See* NRC Report at 10.
[10] *See* Merv Fingas, A REVIEW OF LITERATURE RELATED TO OIL SPILL DISPERSANTS 1997-2008 5 (Sept. 2008), available at http://www.pwsrcac.org/docs/d0053000.pdf.
[11] NRC Report at 207.
[12] *Id.*
[13] The NRC Report, however, suggests that the effect of dispersants on the fur and feathers of animals and birds— *e.g.*, potential negative effects on waterproofing—requires further study. See NRC Report at 196, 274.
[14] See Fingas at 22.
[15] See, e.g., Richard Camilli et al., *Tracking Hydrocarbon Plume Transport and Biodegradation at Deepwater Horizon, SCIENCE* (Aug. 19, 2010), http://www.sciencemag.org/cgi/content/abstract/science.1195223); Terry C. Hazen, et al., *Deep-Sea Plume Enriches Indigenous Oil-Degrading Bacteria, SCIENCE* (Aug. 24, 2010), *http://www.sciencemag.org/cgi/content/abstract/science.1195979.*
[16] *See* Fingas at 22.
[17] 33 U.S.C. § 31 1(d)(2)(G).
[18] Id.
[19] 40 C.F.R. § 300.910(a).
[20] 40 C.F.R. § 300.910(b).
[21] "There is established for each area designated by the President an Area Committee comprised of members appointed by the President from qualified personnel of Federal, State, and local agencies." Clean Water Act § 311(j)(4)(A); 40 C.F.R. § 300.5. In the spill-affected area there are two "areas" (and thus Area Committees) for Louisiana, three for Texas, two for Northwest/West Florida, and one for Mississippi/Alabama.
[22] "On-scene coordinator (OSC) means the federal official pre-designated by EPA or the USCG [U.S. Coast Guard] to coordinate and direct responses under subpart D, or the government official designated by the lead agency to coordinate and direct removal actions under subpart E of the NCP." 40 C. F.R. § 300.5. Rear Admiral Mary Landry was the Federal On-Scene Coordinator until June 1, 2010, when Rear Admiral James Watson assumed that role. On July 12, 2010, Rear Admiral Paul Zukunft replaced Admiral Watson as the Federal On-Scene Coordinator.
[23] Clean Water Act § 311(j)(4)(B); see also 40 C.F.R. § 300.910(a) ("RRTs Area Committees shall address, as part of their planning activities, the desirability of using appropriate dispersants, surface washing agents, surface collecting agents, bioremediation agents, or miscellaneous oil spill control agents listed on the NCP Product Schedule, and the desirability of using appropriate burning agents.").
[24] *See* NRC Report at 3.
[25] *Id.* at 2.
[26] *See* Biello, *Fighting Pollution with Pollution.* Discrepancies between the pre-approval tests and EPA's post-spill toxicity testing results suggest that there were potential flaws in the earlier testing, although it may not be possible to resolve that question definitively at this late date. The pre-approval tests found differences in toxicity between dispersants that did not appear in the EPA test. *Gulf Coast Oil Spill: Small Business and the Cleanup Effort Before the S. Subcomm. On Small Business and Entrepreneurship*, 111th Cong. (June 18, 2010) (statement of Carys Mitchelmore) ("Noteworthy is that the reference toxicant LC50s for the different dispersants listed on the NCPPS differ by orders of magnitude, up to nearly 300-fold. For example, in Table 2 reference toxicant data for the mysid shrimp tests range from an LC50 (ppm, 96-hr) from 0.98 (for Sea Brat #4) to 267.7 (for Nokomis 3-F4). One product (Nokomis 3-AA) used copper sulfate as a reference toxicant instead of the EPA-required SDS reference toxicant. These issues are of concern if you are trying to compare the relative toxicity of the dispersants. Indeed, this currently, cannot be accurately assessed given the data presented on the NCPPS. These toxicity tests should be repeated.").

[27] The Region 6 dispersant guidelines pre-authorize use of any dispersant on the National Product Schedule in water at least ten meters deep and at least three miles from shore. Region 4's dispersant guidelines give the same general pre-authorization, but exclude certain geographic areas. Both sets of guidelines also provide the Federal On-Scene Coordinator with a checklist of factors to consider—not including overall volume or duration—in determining whether to permit dispersant use. RRT-6, FOSC DISPERSANT PRE-APPROVAL GUIDELINES AND CHECKLIST (2001), *available at* http://www.losco.state.la.us/pdf_docs/ RRT6_Dispersant_Preapproval_2001.pdf; REGION IV REGIONAL RESPONSE TEAM RESPONSE AND TECHNOLOGY COMMITTEE DISPERSANT WORKGROUP, USE OF DISPERSANTS IN REGION IV (1996), *Available At* http://ocean.floridamarine.org/acp/ mobacp/PDF/ANNEXES/RRT%20IV%20Dispersant%20Policy.pdf.

[28] Figures on the volume of dispersant use are either taken directly from, or calculated based on data in, the *Operations and Ongoing Response daily reports for the Deepwater Horizon Response*, available at www.restorethegulf.gov [hereinafter Restore the Gulf Estimates].

[29] Campbell Robertson, *White House Takes a Bigger Role in the Oil Spill Cleanup*, N.Y. TIMES (Apr. 29, 2009).

[30] Restore the Gulf Estimates.

[31] *Id.*

[32] *See* Transcript from Press Briefing, Coast Guard Commandant Thad Allen and Assistant to the President for Homeland Security John Brennan (May 1, 2010), *available at* http://coastguardnews.com/gulf-oil- Deepwater Horizon Response, *The Ongoing Administration-Wide Response to the Deepwater Horizon Oil Spill* (May 3, 2010), available at http://www.restorethegulf.gov/release/2010/05/03/ongoing-administration-wide-response -deepwater-horizon-oilspill [hereinafter May 3, 2010 Release].

[33] May 3, 2010 Release.

[34] Interview with Coast Guard official.

[35] *See* H.R. Committee on Energy and Commerce, 111th Cong., CHRONOLOGY OF DEEPWATER HORIZON EVENTS, *available at http://energycommerce.house.gov/documents/20100615/ EE.Attachment.A.2010.6.12.pdf.*

[36] See Conference Call with Lisa P. Jackson, EPA Administrator (May 12, 2010), available at http://www.epa.gov/bpspill/dispersants/may12transcript-final.pdf; see also Joel Achenbach & Steven Mufson, Engineers draw battle lines in effort to plug gulf oil well: *'Top hat,' 'hot tap' among tactics pu*rsued; uncertainties still loom, WASH. POST (May 11, 2010).

[37] *See* EPA, DISPERSANT MONITORING AND ASSESSMENT DIRECTIVE FOR SUBSURFACE DISPERSANT APPLICATION (May 10, 2010); *available at* http://www.epa.gov/bpspill/ dispersants/subsurface-dispersant-directive-final.pdf; EPA, DISPERSANT MONITORING AND ASSESSMENT DIRECTIVE FOR SUBSURFACE DISPERSANT APPLICATION–ADDENDUM 1 (May 14, 2010), available at http://www.epa.gov/bpspill/ dispersants/subsurface-dispersant-addendum-final.pdf; Press Release, Deepwater Horizon Incident Joint Information Center, *Coast Guard and EPA Approve Use of Dispersant Subsea in Further Effort to Prevent Oil From Reaching U.S. Shoreline* (May 15, 2010), available at http://www.restorethegulf.gov/ release/2010/05/15coast-guard-and-eps-approve-use-dispersant-subseafurther-effort-prevent-oil -reac.

[38] Jeff Goodell, *The Poisoning*, ROLLING STONE (July 21, 2010). Administrator Jackson gave a wide-ranging and candid interview to *Rolling Stone*, in which she stated that she had told her aides that the approval decision was among the hardest she had ever made. She also reportedly said that BP argued for subsea application as a method that would reduce the overall volume of chemicals discharged into the marine ecosystem. *Id.*

[39] *See*, e.g., Elizabeth Rosenthal, In Gulf of Mexico, A Huge Experiment with Chemical Dispersants, N.Y. TIMES (May 6, 2010) (characterizing "BP and federal officials [as] engaging in one of the largest and most aggressive experiments with chemical dispersants in the history of the country, and perhaps the world.").

[40] *See* EPA, DISPERSANT MONITORING AND ASSESSMENT DIRECTIVE FOR SUBSURFACE DISPERSANT APPLICATION– ADDENDUM 2 (May 20, 2010), *available at http://www.epa.gov/bpspill/dispersants/directive-addendum2.pdf* [hereinafter EPA ADDENDUM 2]. BP had used Corexit 9500 and 9527, though it discontinued use of the latter early on during the spill because Corexi t 9527 contained 2-butoxyethanol, which had allegedly created health problems for Exxon Valdez workers. *See* Elana Schor, *Ingredients of Controversial Dispersants Used on Gulf Spill Are Secret No More*, N.Y. TIMES (June 9, 2010).

[41] *See* EPA, NATIONAL CONTINGENCY PLAN PRODUCT SCHEDULE TOXICITY AND EFFECTIVENESS SUMMARIES, *available at* http://www.epa.gov/emergencies/content/ncp/tox_tables.htm#dispersants.

[42] *See* EPA ADDENDUM 2. $LD_{50}$ is the dose that is lethal to 50% of the test population. *Menidia* is a genus of silverside fish found in the Gulf of Mexico. *Mysodopsis* are a type of shrimp used for toxicity testing. The reference to toxicity levels "at or below" designated $LD_{50}$ levels was confusing, because a higher $LD_{50}$ actually means a safer substance.

[43] *See id.*

[44] *See* Letter from Douglas J. Suttles of BP to Rear Admiral Mary Landry, Commander, Eighth Coast Guard District, and Samuel Coleman, Director, Superfund Division EPA Region 6 (May 20, 2010), *available at* http://www.epa.gov/bpspill/dispersants/5-21bp-response.pdf [hereinafter Suttles Letter]. This letter refers to the directive (EPA Addendum 2) as having a May 19, 2010 date.

[45] Corexit is a product of the Nalco Company, headquartered in Napier, Illinois. Corexit 9500A contains petroleum distillates, propylene glycol, and a proprietary organic sulfonate (a type of detergent). See Safety Data Sheet, Nalco Company, Corexit EC 9527A (May 11, 2010), *available at* http://www.deepwaterhorizonresponse.com/ posted/2931/Corexit_EC9527A_MSDS.539295.pdf (last visited Sept. 14, 2010).

[46] *See* Suttles Letter.

[47] *See id.* The manufacturer tests were also conducted by different laboratories and on dispersants mixed with No. 2 fuel oil, not Louisiana sweet crude. *See* EPA, DISPERSANTS TOXICITY TESTING–PHASE II QUESTIONS AND ANSWERS (Aug. 2, 2010), *available at* http://www.epa.gov/BPSpill/dispersants/qanda-phase2.pdf [hereinafter DISPERSANTS TOXICITY Q&A] (explaining in answer to question seven that No. 2 fuel oil is not the oil in the Gulf).

[48] *See* Suttles Letter.

[49] Id.

[50] Id.

[51] Id.

[52] See id.

[53] Jackson Press Conference.

[54] Id.

[55] Id.

[56] Id.

[57] Letter from Lisa P. Jackson, EPA Administrator, to David Rainey, V.P. of Gulf of Mexico Exploration, BP Exploration and Production (May 26, 2010), available at *http://www.epa.gov/bpspill/dispersants/Rainey-letter052610.pdf.*

[58] Id.

[59] *See id.*

[60] *See id.*

[61] *Id.*

[62] *See* EPA, DISPERSANT MONITORING AND ASSESSMENT DIRECTIVE FOR SUBSURFACE DISPERSANT APPLICATION– ADDENDUM 3 (May 26, 2010), available at http://www.epa.gov/bpspill/dispersants/directive-addendum3.pdf.

[63] *See* Press Release, Deepwater Horizon Incident Joint Information Center, EPA Releases *First Round of Toxicity Testing Data for Eight Oil Dispersants* (July 1, 2010), *available at*

http://www.deepwaterhorizonresponse.com/go/doc/2931/731363/ The dispersants were Corexit 9500A, Di spersi t SPC 1000, JD-2000, Nokomis 3-AA, Nokomis 3-F4, SAF-RON GOLD, Sea Brat #4 and ZI-400.

[64] *Id.*

[65] *See* Restore the Gulf Estimates.

[66] *Id.*

[67] *Id.*

[68] *Id.* Note that "totals" may not exactly correspond to the sum of components, apparently due to rounding, use of different data sources, or minor calculation errors in the original source. Weekly totals in the table are calculated by subtracting the government figures for cumulative use between weeks; the results do not always correspond to the amounts reported separately for subsea and surface use. The discrepancies in the government figures appear to be relatively minor, however.

[69] The "lower bound" refers to the fact that, according to the source, use was "more than" these amounts.

[70] Letter from Rear Admiral James A. Watson, Federal On-Scene Coordinator, to Regional Response Team, Federal Region VI (June 22, 2010), *available at* http://markey.house.gov/ images *24.pdf.*

[71] *Id.*

[72] *See* notes 94 and 95, infra.

[73] Letter from Peter Gautier, Chief of Staff, National Incident Command, to Rep. Edward J. Markey (July 12, 2010), *available at* http://markey.house.gov/docs/07-30-10ejmtocgdispersants.pdf.

[74] Letter from Rep. Edward J. Markey to Admiral Thad W. Allen, National Incident Commander, United States Coast Guard (July 30, 2010), *available at* http://markey.house.gov/docs/07-30-10ejmtocgdispersants.pdf.

[75] Matthew L. Wald, *Despite Directive, BP Used Dispersant Often,* Panel Finds, N.Y. TIMES (Aug. 1, 2010). Commission staff has learned from EPA staff that this claim was based on a comparison to the highest daily rate of use, rather than a comparison of amounts used on a weekly basis. See Table 1.

[76] Press Conference with Thad Allen, National Incident Commander, U.S. Coast Guard (Aug. 1, 2010), *available at* http://www.restorethegulf.gov/release/2010/08/01/transcript-press-briefing-national-incident-commander-admiralthad-al len-0.

[77] *Allen 'satisfied' with dispersant use in Gulf oil disaster,* CNN (Aug. 2, 2010), http://edition.cnn.com/2010/US/08/01/gulf.oil.spill/index.html#fbid=CAHE69XH089.

[78] *Id.*

[79] 40 C.F.R. §§ 300.145, 300.175, 300.210.

[80] 7 U.S.C. §§ 1531 et seq.

[81] 16 U.S.C. §§ 1361 et seq.

[82] 16 U.S.C. §§ 1801 et seq.

[83] *See supra note* 26; Biel lo, *Fighting Pollution with Pollution* (quoting toxicologist Carys Mitchelmore's statement that "[i]f you think the data on COREXIT is bad, try to find any decent toxicology data on the alternatives").

[84] In interviews with Commission staff, responders stated that EPA representatives on the scene, unlike representatives from other government agencies, were not empowered to make binding decisions notwithstanding EPA's claims of authority over the use of dispersants. Instead, those EPA representatives had to relay information to agency superiors, which inevitably delayed decisions that needed to be made quickly. In addition, these response participants from other federal agencies stated that the EPA on-scene representatives sometimes lacked the necessary experience in oil spill response and that EPA scientists with such experience were not being adequately consulted in EPA's decision-making process. Finally, these individuals expressed an overall concern that EPA's internal decision-making procedures were simply not organized to make the kind of rapid decisions necessary in the

oil spill context, which are quite different from the lengthier deliberative processes that mark the kind of long-term regulatory rulemakings in which EPA more routinely engages.

85 In May, the Regional Response Teams asked for scientific input to direct their future dispersant use and, to that end, fifty experts met together on May 26 and 27, 2010, at Louisiana State University for the "Deepwater Horizon Dispersant Use Meeting." Coastal Response Research Center, DEEPWATER HORIZON DISPERSANT USE MEETING REPORT 5 (June 4, 2010) *available at* http://www.crrc.unh.edu/dwg/dwh_dispersants_use_meeting_report.pdf. In the meeting, the experts split into four breakout groups: (1) efficacy and effectiveness of surface and deep ocean dispersants use; (2) physical transport and chemical behavior of dispersants and dispersed oil; (3) exposure pathways and biological effects resulting from deep ocean application of dispersants; and (4) exposure pathways and biological effects resulting from surface application of dispersants. *Id.*

86 *Id.* at 4.

87 Interviews with government officials.

88 The Unified Command's working relationship with the state of Louisiana on the issue of dispersant use may not have been as cooperative. Commission staff heard in multiple interviews with state and federal government officials that Louisiana disagreed with the Unified Command's decision to apply dispersants subsea.

89 Interviews with government officials. EPA staff has told Commission staff that, while decreasing volatile organic compounds at the surface was not a primary justification for permitting subsea dispersant use, it was an important benefit that became apparent after use commenced, even without conclusive proof of a causal link between subsea use and the reduction.

90 *See* DISPERSANTS TOXICITY Q&A.

91 EPA, COMPARATIVE TOXICITY OF LOUISIANA SWEET CRUDE OIL (LSC) AND CHEMICALLY DISPERSED LSC TO TWO GULF OF MEXICO AQUATIC TEST SPECIES (July 31, 2010), available at http://www.epa.gov/BPSpill/reports/phase2dispersant-toxtest.pdf.

92 Conference Call with Paul T. Anastas, EPA Assistant Administrator (Aug. 2, 2010), available at www.epa.gov/bpspill/dispersants/conference [hereinafter A nastas Conference Call]. Later data supports this conclusion about oxygen levels. See Paul Voosen & Katie Howell, *Gulf Spill Roundup: Subsurface Oil Increasingly Difficult to DetectNOAA, E&E NEWS* (Sept. 2, 2010), http://www.eenews.net/ eenewspm/print/2010/09/07/1.

93 *See* Anastas Conference Call.

94 DEEPWATER HORIZON MC252 GULF INCIDENT OIL BUDGET (Aug. 4, 2010), *available at* http://www.noaanews.noaa.gov/stories2010/PDFs/DeepwaterHorizonOilBudget20100801.pdf; Jane Lubchenco et al., BP DEEPWATER HORIZON OIL BUDGET: WHAT HAPPENED TO THE OIL? (Aug. 4, 2010), *available at* http://www.deepwaterhorizonresponse.com/posted/2931/Oil_Budget_description_8_3_FINAL.844091.pdf.

95 For further discussion, *see* the Draft Staff Working Paper on the amount and fate of the oil.

96 *Id.*

97 The subsequent reports include, ordered by their release date: The Georgia Sea Grant report (Aug. 17, 2010), available at http://uga.edu/aboutUGA/joye_pkit/GeorgiaSeaGrant_OilSpillReport8-16.pdf, which suggested that only a small amount of biodegradation had occurred; a peer-reviewed Woods Hole Oceanographic Institution Team report (Aug. 19, 2010), which suggested the existence of an "oil plume" and that rapid biodegradation might not be occurring (see Cami ll i et al., Tracking Hydrocarbon Plume Transport); a peer-reviewed article published by a team from the Lawrence Berkeley National Laboratory (Aug. 24, 2010), which found evidence of increased microbial respiration within the "plume" and concluded that biodegradation rates were "faster than expected" (*see* Hazen, et al., Deep-Sea Plume); and, most recently, a peer-reviewed report published in Science Express on September 16, 2010 (David L. Valentine et al., Propane Respiration Jump-Starts Microbial Response to a Deep Oil Spill, SCIENCE EXPRESS ), which added another nuance

by suggesting that, while most of the initial degradation was of gaseous hydrocarbons (not liquid oil), this could prime bacteria to degrade other hydrocarbons in the aging plume.

98 DEEPWATER HORIZON MC252 GULF INCIDENT OIL BUDGET (Aug. 4, 2010), available at http://www.noaanews.noaa.gov/stories2010/PDFs/DeepwaterHorizonOilBudget20100801.pdf (concluding that 16% of the oil was naturally dispersed at the wellhead, while only 8% was chemically dispersed subsea or at the surface).

99 For example, as explained by NOAA Administrator Jane Lubchenco, "one of the worst case scenarios involving longer exposures due to dispersed oil—big losses of spawning bluefish tuna populations—may not be detectable for years." See Eli Kintisch, An Audacious Decision in Crisis Gets Cautious Praise, 329 SCIENCE 735, 736 (2010).

100 EPA, UNDERSTANDING OIL SPILLS AND OIL SPILL RESPONSE 13 (1999).

101 NRC Report at 4.

102 *Id.*

103 *Id.* at 5-8; see Elana Schor, Oil Spill Dispersants Shifting Ecosystem Impacts in Gulf, Scientists Warn, N.Y. TIMES (July 30, 2010) ("[Dr. Nancy] Kinner said the National Research Council's report outlined a $40 million plan for dispersant research, but a quarter of the money materialized over the past five years.").

104 *See* CRRC Report at 7 (confirming need for research into "chemical dispersion," including understanding "the long-term fate of chemically dispersed oil" and "investigating...multiple oil (including heavy) and dispersant types.").

105 Memorandum from Admiral Thad. Allen, National Incident Commander, Deepwater Horizon Response, to Rear Admiral Paul Zukunft, Federal On-Scene Coordinator (Aug. 3, 2010), available at http://www.deepwaterhorizonresponse.com/external/content/document/2931/875939/1/SUB-SURFACE%20OIL%20AN D%20D ISPERSA NT%20DETECTION ,%20SAM PL ING%20AN D%20M ON ITORIN G%20STRATEGY.pdf; see also Press Conference with Admiral Thad Allen, National Incident Commander, Deepwater Horizon Response, and Dr. Jane Lubchenco, NOAA Administrator (July 27, 2010), available at http://www.deepwaterhorizonresponse.com/go/doc/2931/829055/.

106 See, e.g., Biello, Fighting Pollution with Pollution.

In: Dispersants in Oil Spills: Issues and Research ISBN: 978-1-61209-241-6
Editors: Edward K. Holm 

*Chapter 2*

# ANALYSIS OF EIGHT OIL SPILL DISPERSANTS USING *IN VITRO* TESTS FOR ENDOCRINE AND OTHER BIOLOGICAL ACTIVITY

***United States Environmental Protection Agency***

## EXECUTIVE SUMMARY

The U.S. Environmental Protection Agency's Office of Research and Development was asked to evaluate the cytotoxicity and potential for interaction with the androgen and estrogen receptors (AR, ER) of eight oil spill dispersants being used, or could be considered for use, in the Gulf of Mexico. These are Corexit 9500 (the current product being used), DISPERSIT SPC 1000, JD 2000, Nokomis 3-F4, Nokomis 3-AA, SAF-RON GOLD, Sea Brat #4, and ZI-400. To address this request, ORD staff and outside collaborators carried out a number of separate studies that were run using *in vitro* (cell-based) assays. A total of 8 cytotoxicity assays, 3 AR agonist assays, 1 AR antagonist assay and 4 ER agonist assays were run on the 8 dispersants, plus reference compounds. Tests were run across a wide range of dispersant concentrations (0.001 to 10,000 parts per million, or ppm). Two dispersants showed a weak signal in one of the four ER assays, but integrating over all of the ER and AR results these data do not indicate that any of the eight dispersants display biologically significant endocrine activity via the androgen or estrogen signaling pathways. All of the dispersants showed cytotoxicity in at least one cell type at concentrations between 10 and 1000 ppm. Both JD 2000 and SAF-RON GOLD tend to be less cytotoxic than the other dispersants. Likewise, DISPERSIT SPC 1000 tends to be more cytotoxic than the other dispersants in the cell-based assays.

*This document has been reviewed in accordance with U.S. Environmental Protection Agency policy and approved for publication. Mention of trade names or commercial products does not constitute endorsement or recommendation for use.*

# CONTRIBUTORS

## U.S. EPA Office of Research and Development

### National Center for Computational Toxicology

David Dix, Keith Houck, Richard Judson, Robert Kavlock, Thomas Knudsen, Stephen Little, Matthew Martin, David Reif, Daniel Rotroff

### National Exposure Research Laboratory

Lora Johnson, Mike Hiatt, Tammy Jones-Lepp, John Zimmerman, Andy Grange, Brian Schumacher, Ed Heithmar

### National Health and Environmental Effects Research Laboratory

Vickie S. Wilson, Phillip Hartig, L. Earl Gray Jr., Christy Lambright, Mary Cardon, Nicola Wrench, Johnathan Furr, Nathan Jeffay, Barbara Collins, Carol Mitchell

## National Institutes of Health

### NIH Chemical Genomics Center,

Menghang Xia, Srilatha Sakamuru, Ruili Huang, Paul Shinn, Christopher P. Austin

# INTRODUCTION / BACKGROUND

The Deepwater Horizon oil spill has led to the use of large amounts of dispersant as part of the integrated approach dealing with the oil spill. Given this fact, questions have arisen about the toxicity of the chemicals used as dispersants themselves. EPA's Office of Research and Development (ORD) was asked to carry out rapid studies to provide information on the potential for toxicity of eight commercially available dispersants. Because some of the dispersants reportedly include nonylphenol ethoxylates (NPEs) that can degrade to isomers of nonylphenol (NP), some of which are proven estrogenic compounds, the potential endocrine effects of the dispersants are of particular concern. For example, NPEs and NPs have been demonstrated to be endocrine disruptors in fish [1]. In response to the request ORD has undertaken a series of short- term *in vitro* studies to determine if any of the dispersants displayed estrogenic, androgenic or other endocrine activity.

ORD developed a strategy to address the questions of endocrine activity and relative toxicity as rapidly as possible. ORD scientists initiated several complementary studies of eight oil spill dispersants being used or considered for use in the Gulf. The issue was to provide some targeted information on the dispersants as quickly as possible. *In vitro* assays are well suited for that purpose. This work complements a study of whole animal toxicity in small fish and brine shrimp also being carried out by ORD. The results of that study are being simultaneously released with this report.

One set of studies used a set of mammalian *in vitro* reporter gene assays in estrogen- responsive and androgen-responsive cells [2, 3] run in-house at ORD laboratories in RTP, NC. Additional studies were conducted by two external labs (NIH Chemical Genomics Center [NCGC] and Attagene Inc.) to run mammalian *in vitro* reporter gene assays to measure androgen and estrogen-response activity. A panel of 74 assays against non-endocrine molecular targets was also included in the Attagene assays. The NCGC and Attagene assays are part of the EPA ToxCast program [4, 5]. All assays evaluated the eight dispersants Corexit® 9500, JD 2000™, DISPERSIT SPC 1000™, Sea Brat #4, Nokomis 3-AA, Nokomis 3-F4, ZI-400 and SAF-RON GOLD. The performance of the assays was characterized by simultaneously running positive and negative control chemicals. Quantitative cytotoxicity measurements were carried out on each of the cell types used. All data analyses and interpretation were carried out by ORD staff.

It is important to note that positive results *in vitro* only demonstrate that a chemical is a potential endocrine disruptor and that follow-up tests will likely be needed in order to refine or confirm the endocrine activity. For example, effects seen *in vitro* may not be expressed *in vivo*, so additional studies would need to be conducted to verify the *in vitro* results and determine if the potential activity was displayed in whole animals and the dosage levels required to affect organisms.

## Project Goals

1. Determine if any of the eight dispersants displayed estrogenic, androgenic or antiandrogenic activity *in vitro* using a variety of well characterized *in vitro* cell-based assays that utilize different approaches for detecting endocrine driven gene expression changes
2. Determine the dispersant concentration that induced cytotoxicity in multiple cell lines and derive an aggregate measure of cytotoxicity that could be use to rank order the chemicals and to compare with *in vivo* toxicity data obtained in aquatic test species.

## Study Summary

One part of the project was carried out by ORD researchers in partnership with the NIH Chemical Genomics Center [NCGC] and Attagene Inc. Two high throughput assay sets were run on the dispersants, a collection of reference chemicals for ER and AR activity, plus nonylphenol compounds. In addition to assays for AR and ER, this phase of the project produced data on a battery of other transcription factor assays which are part of multiplexed panels including AR and ER assays. Cytotoxicity was evaluated in three cell lines over a range of concentrations.

The other phase of the study was carried out in-house by ORD researchers using multiple assays [2, 3] to measure interaction between the eight dispersants plus reference chemicals and ER or AR. In particular, this work evaluated the eight dispersants for estrogen agonist activity in an estrogen-responsive transcriptional activation assays (ER-TA), for androgen agonist activity in two androgen-responsive transcriptional activation assays (AR-TA), MDA-kb2 and CV-1 assays and for androgen antagonist activity in the MDA-kb2 assay in competition with 1 nM Dihydrotestosterone (DHT). Cytotoxicity was evaluated in each assay at every concentration by both a biochemical assay which assessed metabolic perturbation and by a visual assessment of cytopathic effect on cell viability and morphology.

## CHEMICALS

All assays evaluated eight commercially available oil spill dispersants that were obtained directly from the respective manufacturers. EPA chose these eight dispersants from the dispersants listed on the National Contingency Plan Product Schedule based on three criteria: 1) lower toxicity of the dispersant or of the dispersant when mixed with oil; 2) availability of sufficient quantities to respond to the Gulf spill; and 3) immediate availability of samples for testing. These included Corexit® 9500 (Nalco Inc., Sugarland TX), JD 2000™ (GlobeMark Resources Ltd., Atlanta, GA), DISPERSIT SPC 1000™ (U.S. Polychemical Corp., Chestnut Ridge, NY), Sea Brat #4 (Alabaster Corp., Pasadena, TX), Nokomis 3-AA (Mar-Len Supply, Inc., Hayward, CA), Nokomis 3-F4 (Mar-Len Supply, Inc., Hayward, CA), ZI-400 (Z.I. Chemicals, Los Angeles, CA) and SAF-RON GOLD (Sustainable Environmental Technologies, Inc., Mesa, AZ). All are liquid solutions. Further information on the dispersants, including the limited publicly available information on the composition of dispersants is given in **Appendix A.1**. The oil spill dispersants were tested *in vitro* at concentrations ranging from 0.01 to 1000 ppm in water (vol:vol).

The assays run by NCGC and Attagene included reference compounds recommended for validating ER /AR assays by ICCVAM (Interagency Coordination Committee on the Validation of Alternative Methods)[6] and the U.S. EPA[7]. A preliminary set of reference compounds was obtained from stocks at EPA facilities in RTP NC. Subsequently, additional samples were obtained from Sigma-Aldrich (St. Louis MO). Included in the reference chemicals are both straight chain and branched NP isomers and corresponding example NPEs. The reference chemicals are 1 7,, -Trenbolone (10161-33-8), 1 7,, -Estradiol (50-28-2), Atrazine (1912-24-9), Bisphenol A (80-05-7), Butylbenzyl phthalate (85-68-7), Dibutyl phthalate (84-74-2), Flutamide (13311-84-7), Linuron (330-55-2), 4–Nonylphenol (linear) (104-40-5), p,p' – DDE (72-55-9), p,p'- Methoxychlor (72-43-5), Procymidone (32809-16-8), Vinclozolin (50471-44-8), 2,4,5-T (93-76-5), Bicalutamide (90357-06-5), Cyproterone acetate (427-51-0), Genistein (446-72-0), 4- (tert-octyl), Phenol (140-66-9), 4-Hydroxytamoxifen (68392-35-8), 5,-androstan- 1 7,, -ol-3 -one

(521-18-6) and 4-Nonylphenol, (branched) (84852-15-3). The two nonylphenol ethoxylates are Tergitol NP-9 (127087-87-0) and Igepal CO-210 (684 12-54-4). Reference chemicals (powder form) were solubilized in DMSO to a final concentration of 20 mM. Further information, including lot and batch are given in **Appendix A.2.**

In the in-house ORD assays, a 17β-Estradiol (E2; 50-28-2) dose response was included on every plate in the ER-TA assay as a positive control. 4-Nonylphenol (branched) (84852-15-3; Fluka) and 17α-Trenbolone (Osaka Hayashi Pure Chemical Industries Ltd., CAS no. 80657-17- 6, purity 99.9%) were also tested in the estrogen mediated assays. A dihydrotestosterone (DHT; Sigma Chemical; CAS 55206-14-9) dose response was included as a positive control on every plate in the AR-TA assays. The potent androgen, 17 α-Trenbolone, was also tested in the androgen agonist assays. Dosing solutions of dispersants and reference compounds were prepared on-site under observation of a Quality Assurance manager. The assays used in the NHEERL assays have been demonstrated [2, 8] to give appropriate responses to known estrogenic or androgenic compounds.

# RESULTS

More detailed assay protocols and statistical analysis methods can be found in the Appendices, as well as a Quality Assurance (QA) Statement.

## Androgen Receptor Agonist Activity

### *AR Agonist Assay 1 - Multiplexed reporter transcription unit (RTU) assay*

*Method Summary:* This assay is part of a multiplexed reporter gene panel run by Attagene Inc. (RTP, NC), under contract to the U.S. EPA (Contract Number EP-W-07-049). This assay consists of 48 human transcription factor DNA binding sites transfected into the HepG2 human liver hepatoma cell line as previously described[9]. This *trans* assay employs a mammalian one-hybrid assay consisting of an additional 25 RTU library reporting the activity of nuclear receptor (NR) superfamily members. The human ligand-binding domain of each nuclear receptor was expressed as a chimera with the yeast

GAL4 DNA-binding domain that activated in trans a 5XUAS-TATA promoter, which regulated the transcription of a reporter sequence unique to each NR RTU. To ensure the specificity of detection, each individual trans-RTU system including both receptor and reporter gene was separately transfected into suspended cells followed by pooling and plating of the transfected cells prior to screening. The *trans* assay evaluates changes in activities of exogenous, chimeric NR-Gal4 proteins. This particular assay evaluated transcription for the Androgen receptor, and uses the code ATG_AR_TRANS. Additional detail of the method is provided in the Appendix B.1. Concentration-response titration points for each compound were fitted as described in Appendix C. For this analysis, there were either 4 replicates in 16 concentrations, except for SAF-RON GOLD which was only tested in 2 replicates and 8 concentrations.

*Results:* No activity was seen for any of the dispersants

### *AR Agonist Assay 2 - AR beta-lactamase Assay*

*Method Summary:* This assay was run at the NIH Chemical Genomics Center (NCGC; Rockville, MD) in collaboration with EPA as part of the Tox21 collaboration[10]. A betalactamase reporter-gene cell-based assay [GeneBLAzer® AR-UAS-bla-GripTite™ assay developed by Invitrogen] was used to measure AR ligand signaling. AR-UAS-bla-GripTite™ HEK 293 cells (AR *bla* cells) were used with assay medium containing 10% dialyzed FBS, 0.1 mM NEAA and 1 mM sodium pyruvate. The assay was performed in clear bottom black Greiner 1536-well plates. R1881, a synthetic androgen agonist, was used as a positive control in the screen. Library compounds were measured for their ability to either stimulate or inhibit the reporter gene activity. Compounds were screened in a titration series in 1536-well format. The fluorescence intensity (405 nm excitation, 460/530 nm emission) was measured using an EnVision plate reader. Data was normalized relative to R1 881 control (40 nM, 100%, for agonist mode and 10 nM, 0%, for antagonist mode), and DMSO only wells (basal, 0% for agonist mode and -100% for antagonist mode). Additional detail of the method is provided in the Appendix B.2. Concentration-response titration points for each compound were fitted as described in Appendix C. For this analysis, there were 8-10 replicates in 24 concentrations.

*Results:* The only dispersant that showed any activity in any of the AR assays was JD 2000, which was active in both the NCGC ER and AR agonist and antagonist assays in all runs with AC50 values ranging from 100-270 ppm (AR) and 82-120 ppm (ER). There was no apparent cytotoxicity in any of the cell line for JD 2000 (see results below). The EMax values for JD 2000 in all of these assays were significantly greater than the values for positive control chemicals, and in the antagonist assays, this dispersant looked like a "super-activator" rather than an antagonist. All of this data taken together indicates strongly that some non-specific activation is occurring that is independent of ER or AR. We have found previously that compounds identified as promiscuous "super-activators" in multiple beta-lactamase reporter gene assays with a narrow potency range (a <3-fold difference in potency is within the experimental variations of these assays) are mostly auto fluorescent (R Huang, unpublished data). Thus, the activity observed for JD 2000 is likely an artifact of the beta-lactamase assay format. Preliminary results from three additional beta-lactamase assays for non-steroid receptor targets all showed the JD 2000 "super-activation". Considering the totality of the data, we conclude that JD 2000 does not exhibit ER or AR transactivation activity. To further confirm that this JD 2000 activity is non-specific and not due to ER or AR activation, we are running several follow-up assays with NCGC: known antiestrogens and antiandrogens are being used to show that JD 2000 activity is not suppressed; and we will complete our analysis of results for the the three non-steroid receptor beta-lactamase assays are being run with JD 2000 to show that this non-specific activity occurs independent of ER and AR.

### *AR Agonist Assay 3 - MDA-kb2 Androgen-responsive transcriptional activation assay*

*Method Summary:* This assay, run in-house by NHEERL researchers, utilized MDA-kb2 cells[2]. These cells contain endogenous human androgen receptor capable of inducing transcription of an androgen responsive gene (AR-TA). This assay employs a luciferase gene driven by the androgen responsive MMTV promoter which has been stably integrated into the cells. When androgen mimicking compounds (i.e. compounds that act as androgen agonists) are present, these cells produce luciferase in a concentration proportionate to the efficacy of the androgen mimic. Nine concentrations of each dispersant were tested for agonist activity. Each concentration was evaluated in a total of eight replicates (two independent evaluations with four replicates per assay). The first dilution of each sample was a 1:100 dilution (i.e. 0.01 dilution or 10,000 ppm) of the dispersant in cell culture medium followed by eight additional 10-fold serial dilutions. Additional detail of the method is provided in the Appendix F.

*Results:* The ability of the eight dispersants to stimulate luciferase expression in this cell line was compared to DHT. The DHT positive control dose induced luciferase expression in MDA Kb2 cells in a precise and reproducible manner within and among the plates (Figure 1. DHT and 17 -Trenbolone data in MDA Kb2 cells). None of the eight dispersants displayed any potential androgenicity (i.e. did not simulate luciferase induction) at any concentration in the MDA Kb2 cell line (Figure 2 dispersant results in MDA Kb2 cells). In fact, all the dispersants displayed significantly reduced luciferase levels due to cytotoxicity at high dispersant concentrations. The synthetic androgen 17-Trenbolone acted as a full androgen agonist at relatively low concentrations.

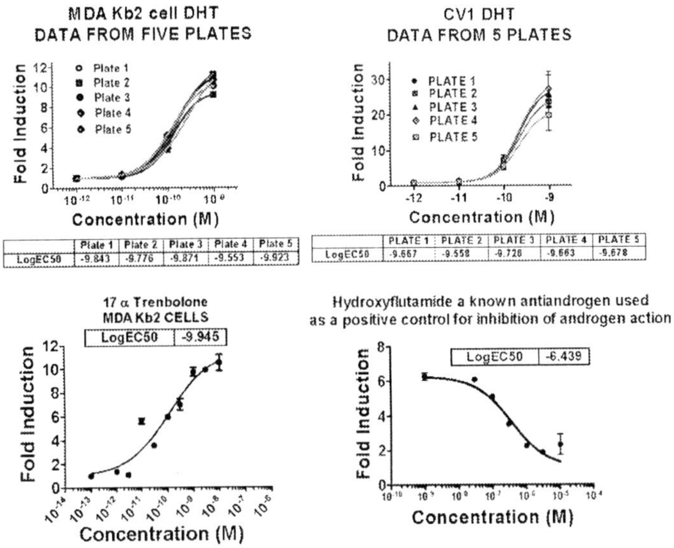


Figure 1. Included are, nonlinear regression plots of the effects of the reference androgen dihydrotestosterone (DHT) in two androgen sensitive cell lines (MDA Kb2 upper left and CV1 upper right), stimulatory effects of the synthetic androgen found in some aquatic systems in MDA Kb2 cells (lower left) and antagonism of the 1 nM DHT by the antiandrogenic drug hydroxyflutamide in MDA Kb2 cells. Data are expressed as fold over the media plus the ethanol control value. The X axis is in log scale. Values are means plus or minus standard errors of the mean.

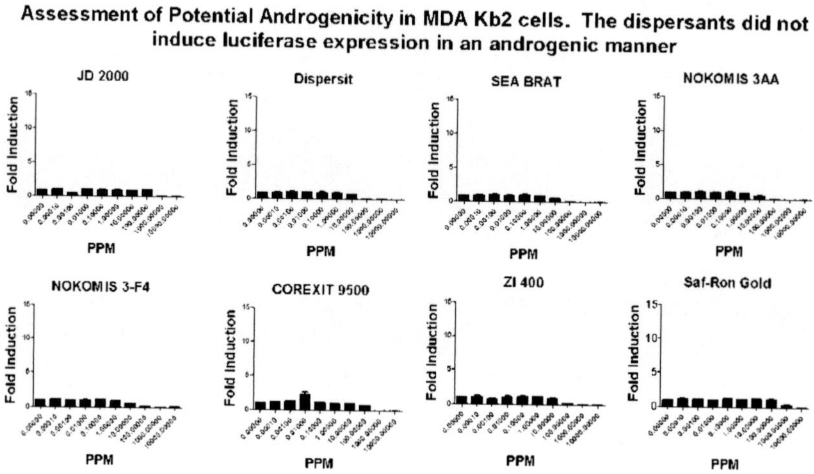


Figure 2. Assessment of the potential androgenic activity of the eight dispersants in MDA Kb2 cells. Data are expressed as fold over the media plus ethanol vehicle control. Values are means plus or minus standard errors of the mean. Dispersants did not stimulate luciferase induction over the control fold value (control fold =1).

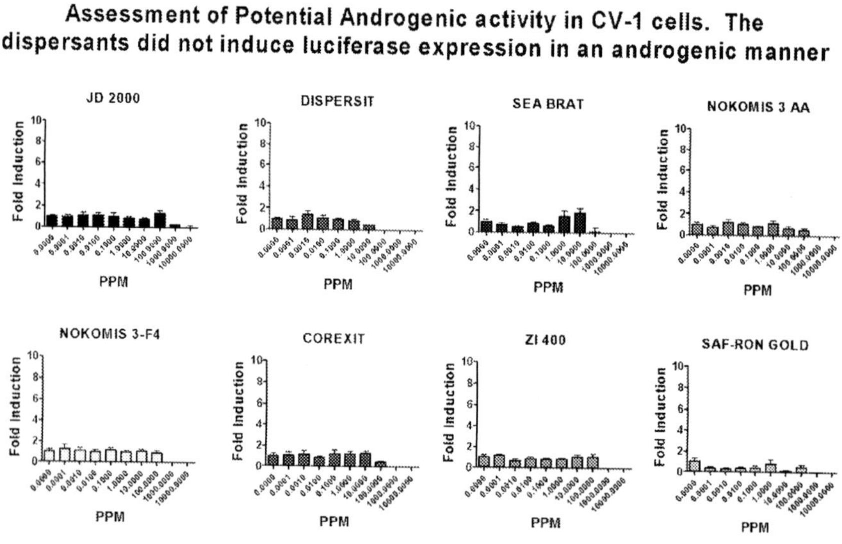


Figure 3. Assessment of the potential androgenic activity of the eight dispersants in CV- 1 cells. Data are expressed as fold over the media plus ethanol vehicle control. Values are means plus or minus standard errors of the mean. Dispersants did not stimulate luciferase induction over the control fold value (control fold =1).

### *AR Agonist Assay 4- CV-1 transient transcription assay*

*Method Summary:* This assay run in-house by NHEERL is similar to the MDA-kb2 in that it also assesses the ability of a compound to mimic an androgen. This assay, however, uses CV-1 cells which do not express either endogenous androgen or estrogen receptors. In contrast to the MDAkb2 assay, both the androgen receptor and the androgen responsive MMTV promoter-luciferase reporter constructs are introduced into the CV-1 cells for each assay via transient transfection. Nine concentrations of each dispersant were tested for agonist activity in both AR-TA assays. Each concentration was tested in quadruplicate. The first dilution of each sample was a 1:100 dilution (i.e. 0.01 dilution or 10,000 ppm) of the dispersant in cell culture medium followed by eight additional 10-fold serial dilutions. Method details are provided in the Appendix F.

*Results:* Similar to the results of the MDA-kb assays, DHT induced precise and reproducible effects on luciferase expression within and among the plates (Figure 1) and none of the eight dispersants displayed any potential androgenicity (i.e. did not simulate luciferase induction) at any concentration in the CV-1 assay (Figure 3). In fact, all the dispersants significantly reduced luciferase level due to cytotoxicity at high concentrations.

## Androgen Receptor Antagonist Activity

### *AR Antagonist Assay 1- MDA-kb2 Androgen-responsive transcriptional activation assay in antagonist mode*

*Method Summary:* The eight dispersants were also evaluated for antagonist activity in the MDAkb2 cell line, run in-house by NHEERL researchers, by testing each dispersant in the presence of a near maximally stimulating concentration dihydrotestosterone (1 nM DHT). In the presence of an anti-androgen, the luciferase activity induced by DHT would be reduced proportionally to the concentration of the anti-androgen. A DHT concentration-response curve was included on each 96-well plate with the dispersants. The well-characterized antiandrogen hydroxyflutamide (CAS 80657-17-6) was run as a positive control (Figure 1). Six concentrations of each dispersant ranging from 0.000 1 ppm to 10 ppm were tested for antagonist activity. Higher concentrations were not evaluated due to cytotoxicity seen in both MTT and CPE assays (discussed later in this document). Each concentration was tested in quadruplicate. Additional detail of the method is provided in the Appendix F.

*Results:* None of the eight dispersants displayed any potential antiandrogenicity (i.e. did not inhibit DHT-induced luciferase induction) at concentrations below 10 ppm (1E-5 dilution). At 10 ppm several of the dispersants reduced DHT induced luciferase activity, but the effects were significant (by ANOVA followed by a post hoc Dunnett's test) for the dispersants SPC 1000 and Nokomis 3-AA (Figure 4). As shown in Figure 5a, these two dispersants were the most toxic of the dispersants to MDA Kb2 cells so it is extremely unlikely that these effects represent competitive inhibition of DHT binding to the ligand binding domain of the androgen receptor. In contrast, hydroxyflutamide, used as a positive control, completely inhibited androgen-induced luciferase induction at concentrations about 1000 fold higher than of the concentration of DHT used in this assay.

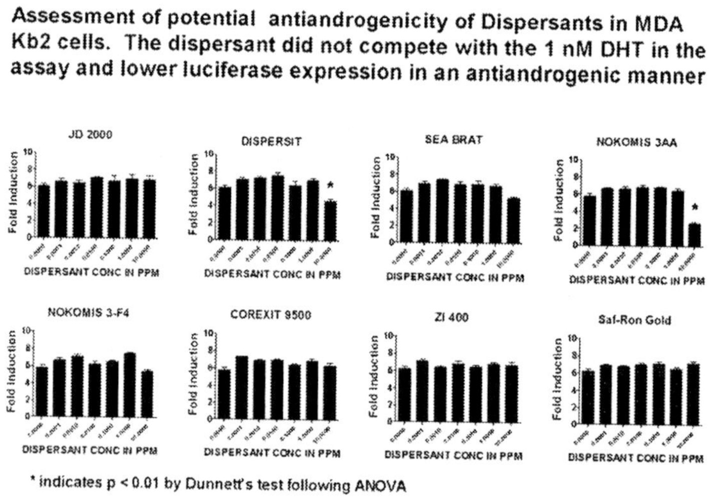


Figure 4. Assessment of the potential antiandrogenic activity of the eight dispersants in MDA Kb2 cells. Data are expressed as fold over the media plus ethanol vehicle control. Values are means plus or minus standard errors of the mean. DISPERSIT SPC 1000 and Nokomis 3-AA were the only dispersants that significantly reduced DHT-stimulated luciferase induction (DHT control fold about 6) an effect which we interpreted to result from cytotoxicity at 10 ppm, the highest concentration used in this assay.

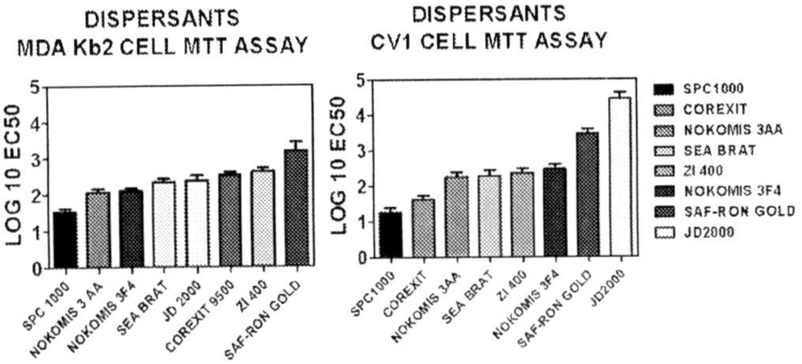


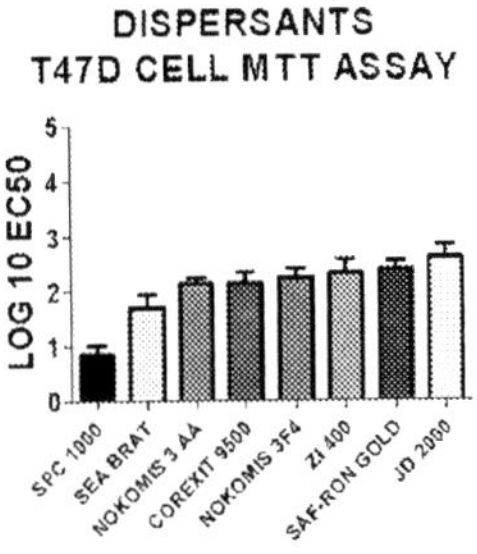


| DISPERSANT | MTT Rank | DISPERSANT | CPE Rank |
|---|---|---|---|
| Dispersit | 1 | Dispersit | 1 |
| Nokomis 3 AA | 2 | Corexit | 2 |
| Seabrat | 3 | JD 2000 | 3 |
| Corexit | 4 | Nokomis 3 aa | 4 |
| Nokomis 3-F4 | 5 | Nokomis 3-F4 | 5 |
| ZI 400 | 6 | Seabrat | 6 |
| JD 2000 | 7 | ZI 400 | 7 |
| Saf-Ron Gold | 8 | Saf-Ron Gold | 8 |

Figure 5. Summary of the Cytotoxic effects of the eight dispersants in the MTT and visual cytopathic examinations (CPE) in three cell lines. The histograms display the EC50 values for a reduction in MTT levels (determined by nonlinear regression on Prism 5.0) for the MDA Kb2 (upper right panel), CV1 (upper right panel) and T47D Kbluc (lower left panel) cells. In the table in the lower right panel, the overall potency of the dispersants in the MTT and CPE assays is shown, with a ranking of 1 being the most potent in inducing cytotoxicity and 8 being the least cytotoxic dispersant. DISPERSIT SPC 1000 was ranked as the most cytotoxic by both methods and SAF-RON GOLD is the least toxic of the eight dispersants. (The appendices contain additional details on this and the EC50 values are compared using a multiple range test).

## Estrogen Receptor Agonist Activity

### *ER Agonist Assay 1- Multiplexed reporter transcription unit (RTU) trans assay*

*Method Summary:* This assay is part of a multiplexed reporter gene panel run by Attagene Inc. (RTP, NC), under contract to the U.S. EPA (Contract Number EP-W-07-049). This assay consists of 48 human transcription factor DNA binding sites transfected into the HepG2 human liver hepatoma cell line as previously described[9]. This *trans* assay employs a mammalian one-hybrid assay consisting of an additional 25 RTU library reporting the activity of nuclear receptor (NR) superfamily members. The human ligand-binding domain of each nuclear receptor was expressed as a chimera with the yeast GAL4 DNA-binding domain that activated in trans a 5XUAS-TATA promoter, which regulated the transcription of a reporter sequence unique to each NR RTU. To ensure the specificity of detection, each individual trans-RTU system including both receptor and reporter gene was separately transfected into suspended cells followed by pooling and plating of the transfected cells prior to screening. The *trans* assay evaluates changes in activities of exogenous, chimeric NR-Gal4 proteins. This particular assay evaluated transcription for the Estrogen receptor alpha, and uses the code ATG_ERa_TRANS. This assay was run in twice in separate weeks, and in each case, run in duplicate. For this analysis, there were either 4 replicates in i6 concentrations, except for SAF-RON GOLD which was only tested in 2 replicates and 8 concentrations. Additional detail of the method is provided in the Appendix B.1. Concentration-response titration points for each compound were fitted as described in Appendix C.

*Results:* We observed statistically significant ER activity in two of the dispersants in the Attagene *trans*-ER assay (Nokomis 3-F4 and ZI-400), detailed in Table 1. Figure 6 (bottom panels) shows the concentration-response curves for the two active dispersants, which have EMax (maximum efficacy) values of between 3 and 4. This is in contrast to 17„-Estradiol (top left panel, blue curve), which has an EMax value of 20. The top right panel of the figure shows the corresponding reference curve for the *cis*-ERE assay, showing that 17„-Estradiol only elicits a response about half of that seen in the *trans* assay. To help interpret these results, we simultaneously analyzed their performance on a set of 19 reference chemicals recommended by ICCVAM[6] and EPA OPPT[7]. This analysis (detailed in Appendix E) shows that these assays perform well for both positive and negative predictive value. The *trans*-ER assay correctly matched ICCVAM expectation for 15 of 17 reference chemicals, with one false positive and one false negative. A comparison of the *cis* and *trans* assays shows that the reference chemicals in the *cis* assay consistently produce EMax values about half of that seen in the trans assay. This would explain the absence of observable activity for these dispersants in the *cis* assay, because we do not consider curves with EMax values below 2. The other curves in the bottom panels of Figure 6 show data for NP and NPE compounds, described below.

**Table 1. Summary results for the Attagene *trans*-ER assay for the positive dispersants**

| Chemical | AC50 (ppm) | EMax | $R^2$ | p-value |
|---|---|---|---|---|
| Nokomis 3-F4 | 16 | 3.9 | 0.65 | 0.00017 |
| ZI-400 | 25 | 3.4 | 0.68 | 0.0041 |

EMax: maximal fold change.

AC50: concentration at which 50 of maximal activity is seen

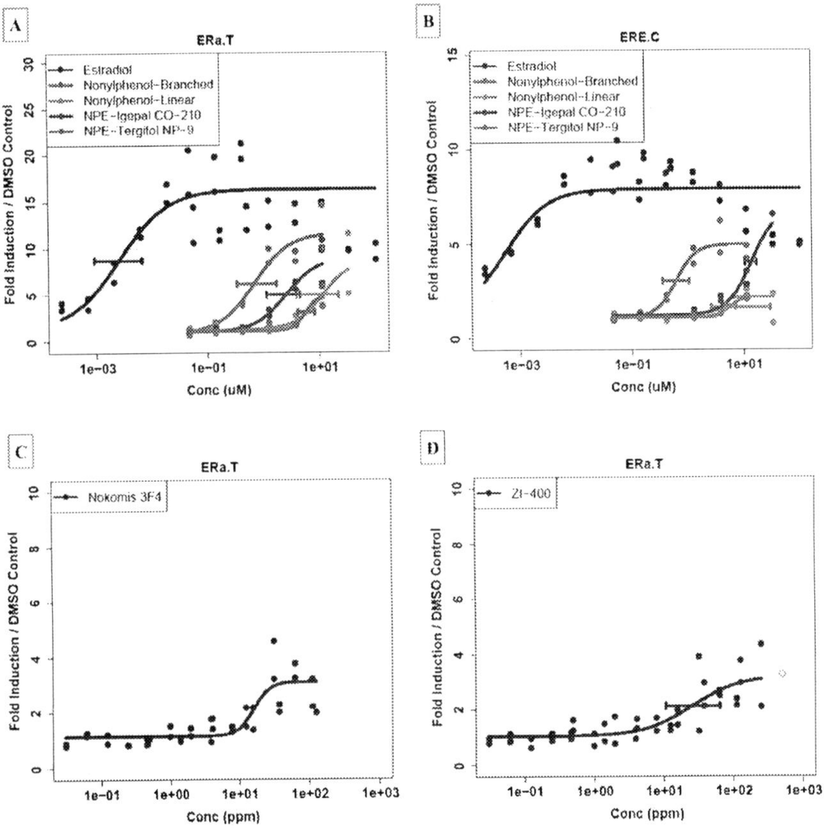


Figure 6. Concentration-response curves for the E2, NP and NPE compounds, and the two dispersants showing activity in Attagene *trans*-ER assay. Top: E2 and the 4 NP / NPE compounds in the Attagene *trans*-ER assay (left) and the Attagene *cis*-ERE assay (right). Bottom: ZI-400 and Nokomis 3-F4 in the Attagene *trans*-ER assay. For the nonylphenol compounds, only two replicates were run.

**Table 2. Results of ER assays on NPs and NPEs**

| Chemical | Assay | AC50 (riM) | $R^2$ | EMax | p-value |
|---|---|---|---|---|---|
| 4-Nonylphenol (linear) 104–40–5 | *trans* -ERa | 11 | 0.77 | 8.3 | 0.29 |
| | *cis*-ERE | 4.3 | 0.55 | 2.7 | 0.096 |
| 4-Nonylphenol (branched) 84852–15–3 | *trans* -ERa | 0.68 | 0.91 | 12 | 0.0049 |
| | *cis*-ERE | 0.61 | 0.092 | 5.4 | 4.9E-5 |
| Tergitol NP-9 127087–87–0 | *trans* -ERa | 5.7 | 0.86 | 4.8 | 0.18 |
| | *cis*-ERE | 5.6 | 0.96 | 2.1 | 0.042 |
| Igepal CO–210 68412–54–4 | *trans* -ERa | 2.5 | 0.89 | 8.5 | 0.19 |
| | *cis*-ERE | 14 | 0.96 | 6.5 | 2.1E-11 |

*Nonylphenol-related activity:* It is known that some of the dispersants contain NPEs. Our initial hypothesis was that any estrogenic activity detected for the complex mixtures could be due to the NPEs or to NP itself generated by *in situ* degradation of the NPE, or residual contamination from synthesis of the NPE. Consequently, we tested two nonylphenols (one linear and one branched, technical grade) and two commercial NPEs in the Attagene assays. Table 2 shows the results of this analysis, and Figure 6 shows the corresponding dose-response curves for the Attagene ER assays. From these data, one can see that these cell-based assays show ER activity for both the NPs and the NPEs. The branched, technical grade NP is the most potent, but the second most potent is the NPE Igepal CO-210. These data indicates that the presence of an NP or NPE in a mixture could give rise to ER activity such as was seen for the dispersants Nokomis 3-F4 and ZI400. Public information (given in Appendix A) indicates that ZI-400 does in fact contain an NPE.

To summarize this section, estrogen receptor (ER) activity was observed in two of the dispersants in the Attagene *trans*-ER assay (ZI-400 and Nokomis 3-F4), although at relatively high concentrations and with low efficacy (EMax). We have also shown that NPs and NPEs are also active in the *trans*-ER assay. Therefore, the activity in ZI-400 and Nokomis 3-F4 is suggestive of the presence of an NP or NPE as part of the mixture. We know that this is the case with ZI-400. The ER effect seen for these dispersants is weak, which is also suggestive of there being only a relatively small amount of NPE or some other estrogenic substance in the total mixture.

### *ER Agonist Assay 2 - Multiplexed reporter transcription unit (RTU) cis assay*

*Method Summary:* This assay is part of a multiplexed reporter gene panel run by Attagene Inc. (RTP, NC), under contract to the U.S. EPA (Contract Number EP-W-07-049). This assay consists of 48 human transcription factor DNA binding sites transfected into the HepG2 human liver hepatoma cell line as previously described[9]. A major difference between the *cis* and *trans* system is that in *cis* activities of endogenous transcription factors are measured. This particular assay evaluated transcription for the Estrogen receptor element (ERE), and uses the code ATG_ERE_CIS. For this analysis, there were either 4 replicates in i6 concentrations, except for SAF-RON GOLD which was only tested in 2 replicates and 8 concentrations. Additional detail of the method is provided in the Appendix B.1. Concentration-response titration points for each compound were fitted as described in Appendix C.

*Results:* No statistically significant activity was seen for any of the dispersants

### *ER Agonist Assay 3 – ER-alpha beta-lactamase Assay*

*Method Summary:* This assay was run at the NIH Chemical Genomics Center (NCGC; Rockville, MD) in collaboration with EPA as part of the Tox21 collaboration[10]. A betalactamase reporter-gene cell-based assay [ER-UAS-bla GripTite$^{TM}$ cell-Based Assay from Invitrogen] was used to measure ER signaling pathway both in agonist and antagonist modes. ER-UAS-bla-GripTiteTM HEK 293 cells (ER *bla* cells) were used with assay medium containing 2% charcoal/dextran treated FBS, 0.1 mM NEAA and 1 mM sodium pyruvate. Cells were cultured in this assay medium overnight in the flasks before the assay. The assay was performed in clear bottom black Greiner 1536-well plates. 17β-estradiol was used as a positive control in the screen. Library compounds were measured for their ability to either stimulate or inhibit the reporter gene activity. Compounds were screened in a titration series in 1536-well format. The fluorescence intensity (405 nm excitation, 460/530 nm emission) was measured using an EnVision plate reader. Data was normalized relative to 17β-estradiol control (20 nM, 100%, for agonist mode and 0.5nM, 0%, for antagonist mode), and DMSO only wells (basal, 0% for agonist mode and -100% for antagonist mode). Concentration-response titration points for each compound were fitted to the Hill equation yielding concentrations of half-maximal stimulation ($EC_{50}$), half-maximal inhibition ($IC_{50}$) and maximal response (efficacy) values. For this analysis, there were 8-10 replicates in 24 concentrations. Additional detail of the method is provided in the Appendix B.3. Concentration-response titration points for each compound were fitted as described in Appendix C.

*Results:* No biologically relevant results were seen for any of the dispersants. See the description above under the corresponding AR assay for JD 2000.

### *ER Agonist Assay 4- T47D-KBluc estrogen-responsive transcriptional activation assay*

*Method Summary:* T47D-KBluc, is an estrogen receptor-mediated transcriptional activation assay (ER-TA) that detects the ability of chemicals to mimic estrogen[8]. This assay was run in- house by NHEERL researchers. The cells contain endogenous human estrogen receptors alpha and beta and are stably integrated with an engineered luciferase reporter gene controlled by triplet estrogen response elements. When the cells are exposed to hormone mimics, the mimicking chemical binds the estrogen receptor and activates production of the luciferase reporter gene. The luciferase product is measured in a light emitting reaction. Additional detail of the method is provided in the Appendix F.

*Results:* The ability of the eight dispersants to stimulate luciferase expression in this cell line was compared to 17β-Estradiol (CAS 50-28-2: a concentration-response curve to E2 was included on each 96 well plate with the dispersants) and to 4-Nonylphenol (branched) (CAS 84852-15-3) (Figure 7 a,b). 17α-Trenbolone (CAS 80657-17-6) was run as a negative control herein (Figure 7d) and as a positive control in the assessment of androgenicity. None of the eight dispersants displayed any potential estrogenicity (i.e. did not simulate luciferase induction) at any concentration in the current investigation (Figure 8). In fact, all the dispersants significantly reduced luciferase levels at high concentrations due to cytotoxicity.

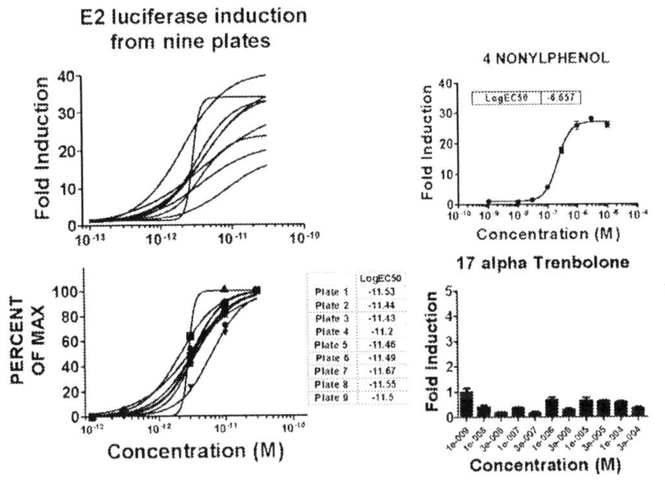


Figure 7. Estradiol 17β-induced (E2) luciferase expression in the nine plates used in the current study with T47D Kbluc cells, expressed as fold over media plus ethanol vehicle control (upper left) and percent of the maximal E2 stimulation (lower left). The effects of the xenoestrogen 4-Nonlyphenol (Branched) are shown in the upper right and the lack of estrogenicity of the synthetic androgen 17α-Trenbolone are shown in the lower right panels. Values are means plus or minus standard errors of the mean.

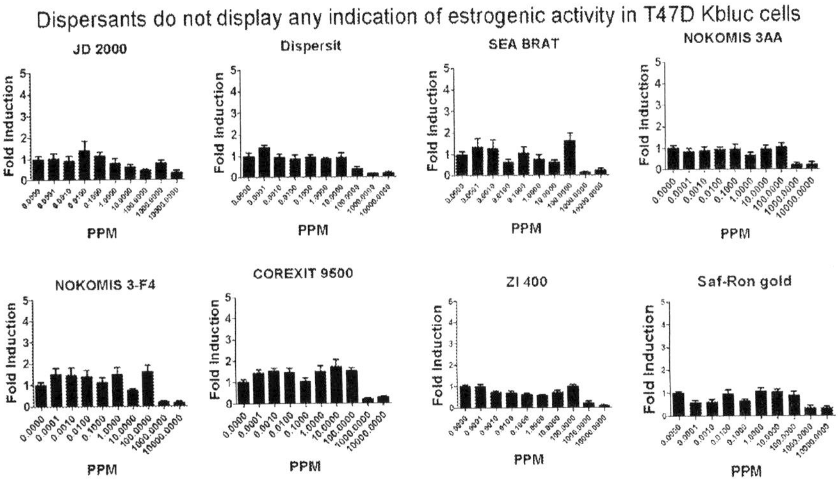


Figure 8. Assessment of the potential estrogenic activity of the eight dispersants in T47D Kbluc cells. Data are expressed as fold over the media plus ethanol vehicle control. Values are means plus or minus standard errors of the mean. Dispersants did not stimulate luciferase induction over the control fold value (control fold =1).

## Cytotoxicity

### *Cytotoxicity Assay 1 –HepG2 Cells*

*Method Summary:* Dispersants were tested for cytotoxicity against HepG2 cells in the MTT (3- (4,5-dimethylthiazol-2-yl)-2,5-diphenyltetrazolium bromide) tetrazolium assay (15) following 24 h chemical exposure to 16 concentrations with an upper concentration of 1000 ppm. All concentrations were run in triplicate. This assay was run by Attagene Inc. LC50 values were determined by fitting curves as described in Appendix C. Results of cytotoxicity assessment are shown below. For this analysis, there were either 4 replicates in 16 concentrations, except for SAF-RON GOLD which was only tested in 2 replicates and 8 concentrations.

### *Cytotoxicity Assay 2 –AR bla Cells*

*Method Summary:* Cell viability after compound treatment was measured in these AR *bla* cells using a luciferase-coupled ATP quantitation assay (CellTiter-Glo viability assay, Promega). This assay was run by the NIH Chemical Genomics Center. The change of intracellular ATP content indicates the number of metabolically competent cells after compound treatment. The cells were dispensed at 2,000 cells/5 L/well for AR *bla* cells in 1,536-well white/solid bottom assay plates using an FRD. The cells were incubated for 5 hrs at 37°C, followed by the addition of compounds using the pin tool. The final concentration range for reference compounds was 11 pM to 92 M, and 0.000144 ppm to 1209.8 ppm for dispersants. The assay plates were incubated for 16 hrs at 37°C, followed by the addition of 5 L/well of CellTiter-Glo reagent. After 30 min incubation at room temperature, the luminescence intensity of the plates was measured using a ViewLux plate reader (PerkinElmer). Data was normalized relative to DMSO only wells (0%), and tetra-n-octylammonium bromide (92 M, -100%). LC50 values were determined by fitting curves as described in Appendix C. Results of cytotoxicity assessment are shown below. For this analysis, there were 8-10 replicates in 24 concentrations.

### *Cytotoxicity Assay 3 –ER bla Cells*

*Method Summary:* Cell viability after compound treatment was measured in these ER *bla* cells using a luciferase-coupled ATP quantitation assay (CellTiter-Glo viability assay, Promega). This assay was run by the NIH Chemical Genomics Center. The change of intracellular ATP content indicates the number of metabolically competent cells after compound treatment. The cells were dispensed at 5,000 cells/5 L/well for ER *bla* cells in 1,536-well white/solid bottom assay plates using an FRD. The cells were incubated a 5 h at 37C, followed by the addition of compounds using the pin tool. The final concentration range for reference compounds was 11 pM to 92 M, and 0.000144 ppm to 1209.8 ppm for dispersants. The assay plates were incubated for 18 hrs at 37C, followed by the addition of 5 L/well of CellTiter-Glo reagent. After 30 min incubation at room temperature, the luminescence intensity of the plates was measured using a ViewLux plate reader (PerkinElmer). Data was normalized relative to DMSO only wells (0%), and tetra-n-octylammonium bromide (92 uM, -100%). LC50 values were determined by fitting curves as described in Appendix C. Results of cytotoxicity assessment are shown below. For this analysis, there were 8-10 replicates in 24 concentrations.

### *Cytotoxicity Results (Assays 1-3)*

Results of Cytotoxicity Assays 1-3 are summarized below the description of Assays 4-6.

### *Cytotoxicity Assays 4 THRU 9: measurements in T47D-KBluc, MDA-kb2, and CV-1 cells (MTT and CPE assessments) (5 independent assessments)*

*Methods summary:* The ability of the dispersants to produce a general toxic effect on each of the cell lines used in the in the NHHERL in-house assays was assessed by both observational and biochemical methods. First, each well of cells in every assay was evaluated by visual microscopic examination utilizing a five point cytopathic effect (CPE) criteria scale ranging from 0 (no visual toxicity) to 4 (total cell death). CPE assessment criteria were as follows: 0 = no observed effect; 1= subtle changes suggesting effect; 2 = definite effects or death in a at least 25% of cells; 3 = 50 to 75% of cells effected; 4 = 100% of cells effected/cell death.

Second, an assessment of the metabolic perturbation of cell health was quantitated by monitoring the ability of cells to metabolize 3-[4,5-Dimethylthiazol-2-yl]-2,5- diphenyltetrazolium bromide (MTT)[1 1]. In this biochemical assay, healthy cells are capable of converting a yellow MTT solution into a blue dye. The healthier the cell the more blue dye produced. This biochemical assay is an indicator that the cells are metabolically active and is a measurement of general cell health. The MTT assay is a quantitative evaluation of mitochondrial function of the cells whereas the first method was a qualitative microscopic cytopathological evaluation (CPE) of cell viability and morphology

### *Cytotoxicity Results (Assays 4-9)*

All eight dispersants disrupted cell function and caused cell death in all three cell lines in the two highest concentrations (0.01 and 0.001, or 10,000 and 1,000 ppm, respectively). Furthermore, none of the dispersants produced any sign of cytotoxicity at concentrations below 1 ppm (Figures 9-12).

**Cytopathological evaluation of dispersant cytotoxicity (eight plates/dispersant, four replicate wells/plate/conc) in MDAKb2, CV 1 and T47D Kbluc cells. Figures are ranked from top left to bottom right in cytotoxic potency at 1, 10 and 100 ppm. At higher concentrations CPE scores were 3-4 for all dispersants and no CPE was observed at concentrations below 1 ppm. The maximum total CPE score is 12, 4 per assay**

| | cytopathic effect: defined |
|---|---|
| no CPE | none observed |
| CPE +1 | subtle changes suggesting effect |
| CPE+2 | definite changes in more than 25 % of cells |
| CPE+3 | 50-75 % of cells effected |
| CPE+4 | 100% cells effected/dead |

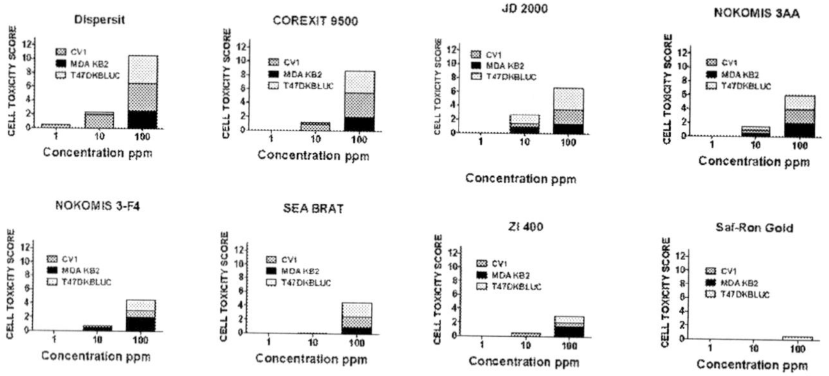


Figure 9.

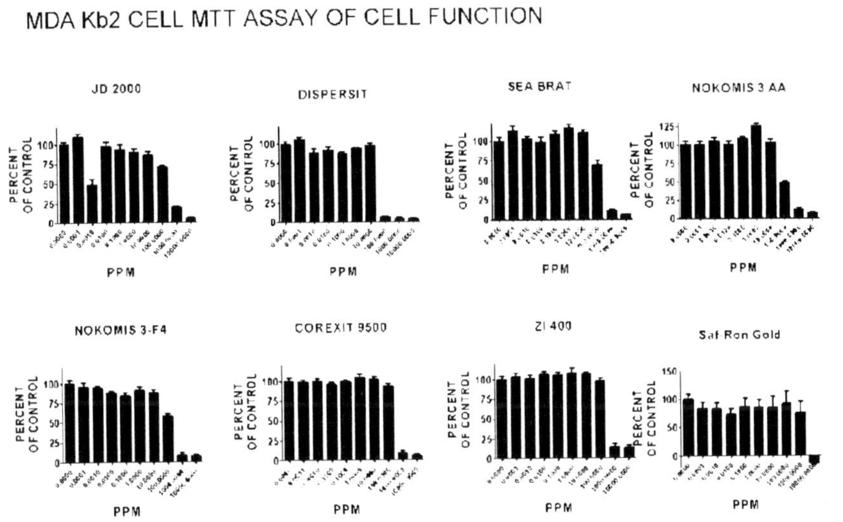


Figure 10. Toxic effects of the eight dispersants on MMT levels in MDA Kb2 cells. A reduction in MTT levels is an indicator of cytotoxcity, seen with all dispersants at the two higher concentrations (1000 and 10,000 ppm) and at 100 ppm with several of the dispersants. Data are expressed as percent of control (media).

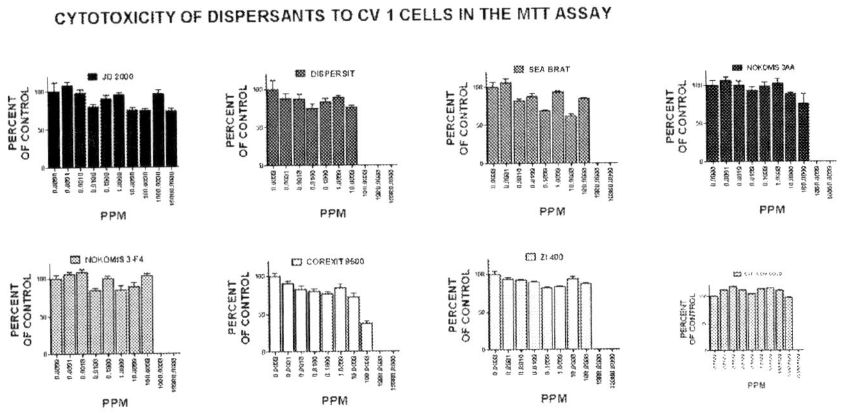


Figure 11. Toxic effects of the eight dispersants on MMT levels in CV- 1 cells. A reduction in MTT levels is an indicator of cytotoxcity, seen with all dispersants at the two higher concentrations (1000 and 10,000 ppm) except JD 2000 (no toxicity was seen at any concentration) and SAF-RON GOLD (toxicity was seen only at the highest concentration of 10,000 ppm). DISPERSIT SPC 1000 and Corexit also induced cytotoxicity at 100. Data are expressed as percent of control (media).

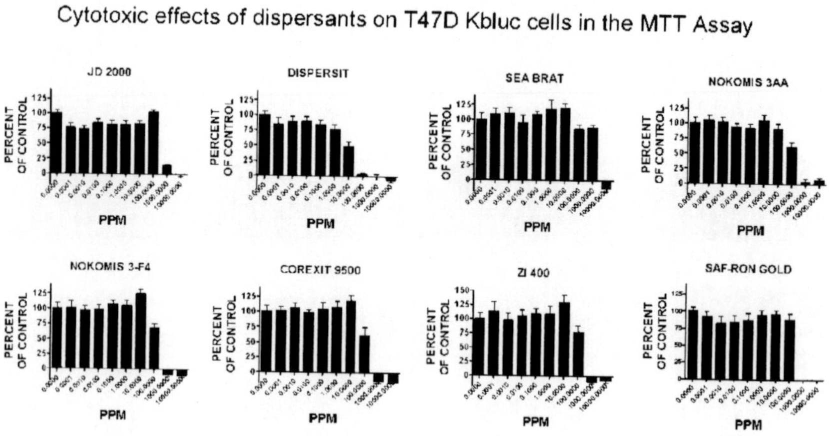


Figure 12. Toxic effects of the eight dispersants on T47D Kbluc cells. A reduction in MTT levels is an indicator of cytotoxicity, seen with all dispersants at the two higher concentrations (1000 and 10,000 ppm) and at 100 ppm with several of the dispersants. Data are expressed as percent of control (media).

The lowest observed effective concentration (LOEC) for dispersant-induced reductions in MTT, estrogen, androgen and antiandrogen assays are reported in Table 3. In the table, the noted changes in the two androgen and the estrogen agonist assays do not result from hormone-like increases in luciferase activity but rather represent significant reductions in luciferase expression that likely result from the cytotoxic effects of the dispersants. Statistical significance was determined using analysis of variance followed by t-tests (LSMEANS) using PROC GLM on SAS 9.1 ($p < 0.01$ was used as the critical value to determine statistical significance).

Since none of the dispersants displayed any effect interpreted as result of the dispersant displaying Endocrine activity we interpret all the results as indications of disruption of cell function and cell death. Since the androgen antagonist assay for antiandrogens did not include the three dispersant highest concentrations (10,000, 1000 and 100 ppm) to avoid most the confounding effects of overt toxicity (seen in the MTT assay with MDA Kb2 cells), the highest concentration in this assay was 10 ppm Dispersants that did not reduce luciferase expression in this assay at any concentration were scored as >10 ppm. In spite of this precaution, the two most cytotoxic dispersants still reduced luciferase expression in this assay, an effect we attribute to less overt cell toxicity.

**Table 3. Summary table of the Lowest Observed Effect Concentration (LOEC) of the eight dispersants in the MTT cytotoxicity in three cell lines, the estrogen agonist assay in T47D Kbluc cells, the agonist assays in CV-1 and MDA K2 cells and the antagonist assay in MDA Kb2 cells**

| Dispersant | MDA Kb2 cells | | | CV1 cells | | T47D Kbluc cells | |
|---|---|---|---|---|---|---|---|
| | MTT Cytotoxicity | Androgen Antagonist Assay*** | Androgen Agonist Assay | MTT Cytotoxicity | Androgen Agonist Assay | MTT Cytotoxicity | Estrogen Agonist Assay |
| | LOEC* (ppm) | LOEC (ppm) | LOEC (ppm) | LOEC (ppm) | LOEC (ppm) | LOEC (ppm) | LOEC (ppm) |
| JD 2000 | 100 | >10 | 1,000 | 10,000** | 1,000 | 1,000** | 10,000** |
| DISPERSIT SPC 1000 | 100 | 10 | 100 | 10 | 100 | 10 | 100 |
| Sea Brat #4 | 100 | >10 | 100 | 1,000** | 1,000 | 10,000 | 1,000 |
| Nokomis 3-AA | 100 | 10 | 10 | 1,000 | 1,000 | 100 | 1,000 |
| Nokomis 3-F4 | 100 | >10 | 10 | 1,000 | 1,000 | 1,000 | 1,000 |
| Corexit 9500 | 1,000 | >10 | 1,000 | 100 | 1,000 | 100 | 1,000 |
| ZI-400 | 1,000 | >10 | 100 | 1,000 | 1,000 | 1,000 | 1,000 |
| SAF-RON GOLD | 10,000 | >10 | 1,000 | 10,000 | 1,000 | 1,000 | 10,000 |

*LOEC (ppm) represents the lowest concentration at which the dispersant consistently reduced the MTT value. Statistical significance was using $p<0.01$ as determined using LSMEANS option of PROC GLM available on SAS 9.1.

** LOEC concentration was equivocal (nonmonotonic response)

*** Antagonist assay for antiandrogens was not run the three highest concentrations (10,000, 1000 and 100 ppm) to avoid most the confounding effects of cell death. The highest concentration was 10 ppm, Dispersants that did not reduce luciferase expression in this assay at any concentration were scored as >10 ppm.

The $EC_{50}$ values for the dispersant dose response curves were determined using nonlinear regression procedures with GraphPad Prism 5.0 software (Figure 5 a,b,c). Ranking the eight dispersants in order of highest to lowest potency in the MTT assays and the CPE assessment in three cells lines indicates that there are some consistent differences among dispersants in their ability to disrupt the function and viability of these cell lines (Figure 5 d). Dispersant SPC 1000 appears to be more toxic in both MTT (below) and CPE assessments.

## Cytotoxicity Summary

For comparison to all of the *in vitro* cytotoxicity assays, we also include LC50 values from whole animal, aquatic species lethality assays for the mysid, *Americamysis bahia*, in a 48-hr static acute toxicity test and an inland silverside, *Menidia beryllina*, 96-hr static acute toxicity test[ 12]. All LC50 values are plotted in Figure 13 and the numerical values are listed in Appendix D. One can see that the cell-based LC50 values overall vary by about two orders of magnitude, and that the values for any given chemical span about one order of magnitude. The rank order of cytotoxicity varied between the various cell types, a not unexpected finding [13]. There is overlap in the range of cytotoxicity for all of the dispersants.

In order to assess, statistically, differential cytotoxicity across the eight dispersants we performed an ANOVA to determine pairwise if any two dispersants were more cytotoxic than the other. We performed this statistical test with and without multiple test correction (Bonferroni). For any dispersant and assay combination that did not achieve an LC50, a default value of 3000 ppm was used; three-fold higher than the highest concentration tested in the relevant assays. LC50 values greater than 3000 ppm were also set to this default value to prevent large extrapolated LC50 values from biasing the results. All six cell-based quantitative cytotoxicity assays were used for this analysis. The resulting p-values, raw and corrected, are provided in Table 4. Both JD 2000 and SAFRON GOLD tend to be less cytotoxic than the other dispersants. Likewise, DISPERSIT SPC 1000 tends to be more cytotoxic than the other dispersants in the cell-based assays.

The aquatic species LC50 values are almost always lower than the cell-based LC50 values. As with the cell-based assays, JD 2000 is the least toxic in the whole animal assay.

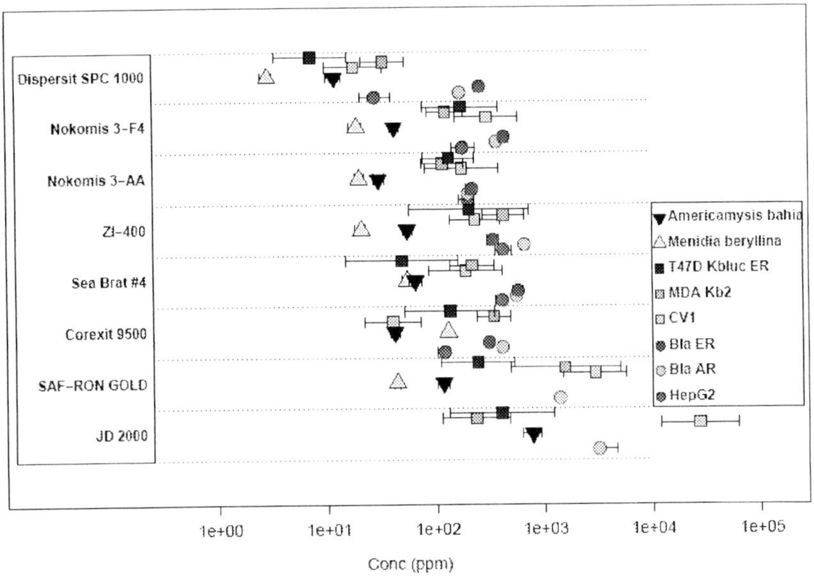


Figure 13. Toxicity data for the dispersants, combining data from cell-based assays in this report with data on aquatic species from a concurrent EPA report [12]. Each horizontal band shows the data for one dispersant. Results are presented from all 6 quantitative cytotoxicity assays. Cell-based LC50 values (concentration at which 50% lethality or effect is observed) are indicated by circles and squares. Aquatic species LC50 values are indicated by triangles. Note that all dispersants were tested in all assays, and missing data points indicate that no toxicity was seen in that assay at the highest concentration tested. 95% confidence intervals are shown for all assays.

## Other Molecular Targets

In addition to ER and AR, we also analyzed the chemical collection (dispersants plus reference chemicals) using a multiplexed reporter gene assay battery that evaluates activity against a panel of transcription factors including nuclear receptors[5, 9]. These assays were run by Attagene Inc. These data also provide a measure of quality control related to the specificity of any endocrine-related activity caused by the dispersants. The description of the assay and a complete list of targets is given in Appendix B.2. All of these assays were carried out twice, one week apart, and in each week, duplicate runs were performed. Figure 14 summarize all of the results for the dispersants. This plot helps illustrate several key points about the data.

**Table 4. Statistical comparison of LC50 cytotoxicity values from cell-based assays across the eight dispersants. All dispersants combinations with a p-value less <0.05 are shaded pink. All values below the diagonal are raw p-values derived from the ANOVA, while all values above the diagonal swere adjusted for multiple testing**

| | JD 2000 | Dispersit SPC 1000 | Sea Brat 4 | Nokomis 3-AA | Nokomis 3-F4 | Corexit 9500 | ZI-400 | SAF-RON GOLD | Bonferroni Corrected P-Value |
|---|---|---|---|---|---|---|---|---|---|
| JD 2000 | | 0.1456 | 0.308 | 0.1876 | 0.2464 | 0.224 | 0.364 | 1 | |
| Dispersit SPC 1000 | 0.0052 | | 0.84 | 1 | 0.644 | 1 | 0.126 | 0.056 | |
| Sea Brat 4 | 0.011 | 0.03 | | 1 | 1 | 1 | 1 | 0.1456 | |
| Nokomis 3-AA | 0.0067 | 0.082 | 0.11 | | 1 | 1 | 0.42 | 0.0756 | |
| Nokomis 3-F4 | 0.0088 | 0.023 | 0.5 | 0.12 | | 1 | 1 | 0.1036 | |
| Corexit 9500 | 0.008 | 0.086 | 0.34 | 0.42 | 0.65 | | 1 | 0.0952 | |
| ZI-400 | 0.013 | 0.0045 | 0.68 | 0.015 | 0.19 | 0.12 | | 0.1652 | |
| SAF-RON GOLD | 0.92 | 0.002 | 0.0052 | 0.0027 | 0.0037 | 0.0034 | 0.0059 | | |
| Raw P-Value | | | | | | | | | |

First, as the concentration of a chemical approaches the cytotoxic level, generalized cell stress occurs, accompanied by broad misregulation of transcription. When this threshold is reached, many assays in this system simultaneously activate, but this activity is assumed to be non-specific. One sentinel of this cell stress behavior is NRF2, which is an indicator of generalized oxidative stress. Therefore, if we see many assays become active at about the same concentration, especially if NRF2 is among them, we can discount any target specificity above that concentration. We see this behavior for Corexit 9500 (~50 ppm), JD 2000 (~500 ppm), Nokomis 3-AA (~75 ppm), Nokomis 3-F4 (~75 ppm), Sea Brat #4 (~90 ppm) and ZI-400 (~50 ppm).

The ER activity for Nokomis 3-F4 occurs at a concentration well below where this nonspecific behavior is indicated. For ZI-400, the confidence intervals for ER and NRF2 overlap, indicating a possibility that the ER result is non-specific.

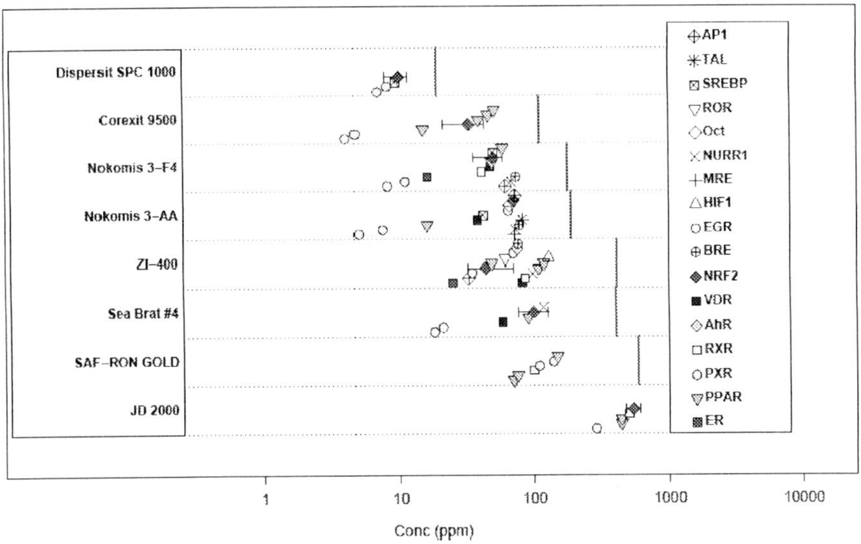


Figure 14. Summary plot of all Attagene *cis* and *trans* assays for dispersants with AC50 values below cytotoxicity levels. Each horizontal band displays data for a single dispersant. The x-value is the AC50. Points are staggered in the y-direction to make overlapping points visible. Where there were multiple assays for a given gene target (e.g. PPAR, PPAR, PPAR) were given a single symbol. For illustration, 95% confidence intervals are shown on assays for NRF2. The vertical red lines indicate the LC50 for cytotoxicity in the HepG2 cells. Dispersants are ordered by decreasing cytotoxicity LC50 values.

The lowest activity that is generally seen is for PXR (Pregnane-X-receptor), which is a xenosensor. This behavior is entirely expected, is common across many classes of organic chemicals, and is not in itself an indicator of toxicity. PXR has been reported to be a xenosensor that acts to protect against endocrine active chemicals[14]. PPAR (peroxisome proliferator activating receptor[ 15-19]) activity is observed for a number of the dispersants, at higher concentrations than is seen for the PXR assays. There is an extensive literature on PPAR activity associated with disease in rodents, although the human relevance is unclear [15-18, 20-23]. However, only for Corexit 9500 and Nokomis 3-AA (and potentially for SAF-RON GOLD) is the PPAR signal well below the level of non-specific activity. Vitamin D receptor (VDR) activity is seen for Sea Brat #4 and Nokomis 3-AA below but near the concentration of nonspecific behavior.

The activity of JD 2000 cannot necessarily be dismissed as being all non-specific, despite it occurring at the same concentration as NRF2 activity. This is because there are only two target families being activated – PXR and PPAR. A similar observation can be made about DISPERSIT SPC 1000. At the concentration of NRF2 activity, we only see activation of two PXR assays and one for SREBP (SREBF1 sterol regulatory element binding transcription factor 1) which is involved in fatty acid synthesis regulation.

The largest effect (in terms of EMax) of any dispersant and assays is for ZI-400 and AhR (Aryl hydrocarbon receptor), with EMax >30. The AhR is well-known for its role in mediating the adaptive metabolism of xenobiotics, and also in the toxicity that follows exposure to 2,3,7,8- tetrachlorodibenzo-p-dioxin (dioxin). This indicates the potential for the presence of a dioxin- like compound, which would be cause for concern. In the ToxCast Phase I data set[4, 5] of 309 chemicals, we saw only three with AhR efficacy higher than is seen with ZI-400. It is not clear though that this effect is specific, given that it occurs in the same concentration range as activity in a number of other targets, and above the NRF2 AC50.

The major conclusions of this section are that several of the dispersants display PXR and PPAR activity at concentrations below where cell stress and cytotoxicity occur. These are expected responses in hepatocytes to xenobiotics. The ordering of dispersants by lowest concentration at which bioactivity occurs is consistent with the ordering based on cytotoxicity. One observation of more general interest is that we are able to detect specific target-based bioactivity in complex mixtures such as these. This is observation is relevant to the challenges of real world chemical toxicity testing, wherein humans and other organisms are often exposed to complex mixtures rather than the pure single compounds that are the subject of typical toxicity testing.

## CONCLUSION

The primary conclusions are as follows:

For six of the eight dispersants tested we found no evidence that they would be capable of interacting with estrogen or androgen receptor function from testing in multiple *in vitro* systems. For the other two dispersants, there was a weak ER signal in one assay. However, integrating over all of the ER and AR results, these data do not indicate that any of the eight dispersants will display biologically significant endocrine activity via the androgen or estrogen signaling pathways. As mentioned previously, NPEs (and their breakdown product NPs) can be endocrine disruptors in fish[1], so the risk of using NPE-containing dispersants should be carefully weighed against the expected benefits. One limitation of the present study is that there are other routes by which chemicals can cause endocrine disruption, as well as other types of toxicity that have not been tested for here. Most importantly though, there were no indications of estrogenic activity for Corexit 9500, the dispersant currently being used in the Gulf of Mexico.

All of the dispersants showed cytotoxicity in at least one cell type at concentrations between 10 and 1000 ppm. Both JD 2000 and SAF-RON GOLD tend to be less cytotoxic than the other dispersants. Likewise, DISPERSIT SPC 1000 tends to be more cytotoxic than the other dispersants in the cell-based assays. The aquatic species LC50 values tend to be lower than the cell-based LC50 values. As with the cell-based assays, JD 2000 is the least toxic in the whole animal assay.

## Supplementary Information

Supplemental information, including a QA Statement, is included in the referenced Appendices.

## QA Summary

All research described in this report was conducted under a comprehensive and rigorous program of quality assurance (QA), as documented in the QA supplemental file. The overall goal of the QA program was to ensure research data were of known and acceptable quality. QA staff surveillance of critical research activities was an important feature of the overall QA approach and ensured quick and effective resolution of any problems. The conclusion of the QA review process is that results presented in this report accurately reflect the raw data obtained during the course of the research and are scientifically valid and defensible.

# References

[1] Jobling, S; Sheahan, D; Osborne, JA; Matthiessen, P; Sumpter, JP. Inhibition of testicular growth in rainbow trout (Oncorhynchus mykiss)

exposed to estrogenic alkylphenolic chemicals. *Environmetnal Toxicology and Chemistry,* 1996, 15, (2), 194-202.

[2] Wilson, VS; Bobseine, K; Lambright, CR; Gray, LE., Jr., A novel cell line, MDA-kb2, that stably expresses an androgen- and glucocorticoid-responsive reporter for the detection of hormone receptor agonists and antagonists. *Toxicol Sci,* 2002, 66(1), 69-81.

[3] Wilson, VS; Cardon, MC; Thornton, J; Korte, JJ; Ankley, GT; Welch, J; Gray, LE; Jr; Hartig, PC. Cloning and in vitro expression and characterization of the androgen receptor and isolation of estrogen receptor alpha from the fathead Minnow (Pimephales promelas). *Environ Sci Technol,* 2004, 38, (23), 63 14-21.

[4] Judson, RS; Houck, KA; Kavlock, RJ; Knudsen, KB; Martin, MT; Mortensen, HM; Reif, DM; Richard, AM; Rotroff, DM; Shah, I; Dix, DJ. Predictive In Vitro Screening of Environmental Chemicals - The ToxCast Project. *Environ Health Perspect,* 2010, 118(4), 485-492.

[5] Martin, MT; Dix, DJ; Judson, RS; Kavlock, RJ; Reif, DM; Richard, AM; Rotroff, DM; Romanov, S; Medvedev, A; Poltoratskaya, N; Gambarian, M; Moeser, M; Makarov, SS; Houck, KA. Impact of environmental chemicals on key transcription regulators and correlation to toxicity end points within EPA's ToxCast program. *Chem Res Toxicol,* 2010, 23(3), 578-90.

[6] ICCVAM ICCVAM Evaluation of In Vitro Test Methods for Detecting Potential Endocrine Disruptors: Estrogen Receptor and Androgen Receptor Binding and Transcriptional Activation Assays. http://iccvam.niehs.nih.gov/docs/endo docs/EDAddendFinal.pdf (15 April),

[7] U.S. EPA OCSPP Harmonized Test Guidelines Series 890 - Endocrine Disruptor Screening Program Test Guidelines. http://www.epa.gov/ocspp/pubs/frs/publications/Test_Guidelines/series890.htm (11- June),

[8] Wilson, VS; Bobseine, K; Gray, LE. Jr., Development and characterization of a cell line that stably expresses an estrogen-responsive luciferase reporter for the detection of estrogen receptor agonist and antagonists. *Toxicol Sci,* 2004, 81(1), 69-77.

[9] Romanov, S; Medvedev, A; Gambarian, M; Poltoratskaya, N; Moeser, M; Medvedeva, L; Diatchenko, L; Makarov, S., Homogeneous reporter system enables quantitative functional assessment of multiple transcription factors. *Nat Methods,* 2008, 5(3), 253-60.

[10] Collins, FS; Gray, GM; Bucher, JR., Toxicology. Transforming environmental health protection. *Science,* 2008, 319(5865), 906-7.

[11] Kelce, WR. Buthionine sulfoximine protects the viability of adult rat

Leydig cells exposed to ethane dimethanesulfonate. *Toxicol Appl Pharmacol,* 1994, 125(2), 237- 46.

[12] Hemmer, MJ; Barron, MG; Greene, RM. *Comparative Toxicity of Eight Oil Dispersant Products on Standard Aquatic Species.* TBD (30 June),

[13] Xia, M; Huang, R; Witt, KL; Southall, N; Fostel, J; Cho, MH; Jadhav, A; Smith, CS; Inglese, J; Portier, CJ; Tice, RR; Austin, CP. Compound cytotoxicity profiling using quantitative high-throughput screening. *Environ Health Perspect,* 2008, 116(3), 284-91.

[14] Kretschmer, XC; Baldwin, WS. CAR and PXR: xenosensors of endocrine disrupters? *Chem Biol Interact,* 2005, 155(3), 111-28.

[15] Guyton, KZ; Chiu, WA; Bateson, TF; Jinot, J; Scott, CS; Brown, RC; Caldwell, JC. A reexamination of the PPAR-a activation mode of action as a basis for assessing human cancer risks of environmental contaminants. *Environ Health Perspect* 2009, *doi:*10.1289/*ehp*.0900758.

[16] Klaunig, JE; Babich, MA; Baetcke, KP; Cook, JC; Corton, JC; David, RM; DeLuca, JG; Lai, DY; McKee, RH; Peters, JM; Roberts, RA; Fenner-Crisp, PA. PPARalpha agonist-induced rodent tumors: modes of action and human relevance. *Crit Rev Toxicol,* 2003, 33(6), 655-780.

[17] Lai, DY. Rodent carcinogenicity of peroxisome proliferators and issues on human relevance. *J Environ Sci Health C Environ Carcinog Ecotoxicol Rev,* 2004, 22(1), 37-55.

[18] Peraza, MA; Burdick, AD; Marin, HE; Gonzalez, FJ; Peters, JM. The Toxicology of Ligands for Peroxisome Proliferator-Activated Receptors (PPAR). *Toxicol. Sci.,* 2006, 90(2), 269-295.

[19] Takeuchi, S; Matsuda, T; Kobayashi, S; Takahashi, T; Kojima, H. In vitro screening of 200 pesticides for agonistic activity via mouse peroxisome proliferator-activated receptor (PPAR)alpha and PPARgamma and quantitative analysis of in vivo induction pathway. *Toxicol Appl Pharmacol,* 2006, 217, (3), 235-44.

[20] Melnick, RL. Is peroxisome proliferation an obligatory precursor step in the carcinogenicity of di(2-ethylhexyl)phthalate (DEHP)? *Environ Health Perspect,* 2001, 109, (5), 437-42.

[21] Peters, JM. Mechanistic Evaluation of PPARa-Mediated Hepatocarcinogenesis: Are We There Yet? *Tox. Sci.,* 2008, 101(1), 1-3.

[22] Peters, JM; Cheung, C; Gonzalez, FJ. Peroxisome proliferator-activated receptor-alpha and liver cancer: where do we stand? *J Mol Med* 2005, 83(10), 774-85.

[23] Poole, TM; Drinkwater, NR. Strain dependent effects of sex hormones on hepatocarcinogenesis in mice. *Carcinogenesis,* 1996, 17(2), 191-6.

In: Dispersants in Oil Spills: Issues and Research ISBN: 978-1-61209-241-6
Editors: Edward K. Holm 

*Chapter 3*

# COMPARATIVE TOXICITY OF LOUISIANA SWEET CRUDE OIL (LSC) AND CHEMICALLY DISPERSED LSC TO TWO GULF OF MEXICO AQUATIC TEST SPECIES

***United States Environmental Protection Agency***

## 1. INTRODUCTION

An estimated 4.9 million barrels of South Louisiana sweet crude (LSC) oil was released into the northern Gulf of Mexico between April 20 and July 15, 2010 caused by the explosion and collapse of the Deepwater Horizon oil exploration platform. The use of dispersants in oil spill response involves tradeoffs between effects to the shoreline and effects to pelagic and deep sea environments. In an effort to mitigate the impact of floating oil on sensitive shoreline habitats the decision was made to use dispersants listed on the U.S. Environmental Protection Agency's (EPA) National Contingency Plan (NCP) Product Schedule (EPA 2010a). Dispersants were applied offshore on the surface as well as underwater at the source of the leak. In an effort to assess options to address this unprecedented event, the EPA conducted independent studies with eight dispersants on the NCP Product Schedule to assess the relative acute toxicity of each dispersant alone, LSC oil alone and dispersant-LSC oil mixtures.

Toxicity testing to determine the hazards of eight dispersant products has been performed in two phases. Phase one included acute toxicity tests with two Gulf of Mexico aquatic species, and *in vitro* cytotoxicity and endocrine screening assays using human cell lines. These tests were performed with each of the dispersants in the absence of oil. The results of the aquatic toxicity tests generally classified the dispersants as ranging from slightly toxic to practically non-toxic to both test species with the exception of one dispersant found to be moderately toxic to fish (Hemmer et al. 2010). The results of the *in vitro* tests showed that cytotoxicity was only observed at concentrations above 10 ppm, and none of the eight dispersants displayed biologically significant estrogenic or androgenic activity (Judson et al., 2010). Results of the *in vitro* tests were similar to the ecotoxicology tests showing generally low dispersant toxicity.

This report summarizes results of the second phase of testing obtained from acute toxicity tests conducted with LSC only and chemically dispersed LSC using each of eight commercial oil dispersants. The same eight dispersants were tested in both Phase One and Two: Corexit 9500A, Dispersit SPC 1000, JD-2000, Nokomis 3-AA, Nokomis 3-F4, Saf-Ron Gold, Sea Brat #4 and ZI-400. The approach utilized consistent test methodologies within a single laboratory which provided a means to independently assess acute toxicity estimates between LSC and dispersantLSC mixtures for all eight dispersants.

# 2. Test Methods

## 2.1. Study Design

In this study the acute toxicity of LSC and dispersant-LSC mixtures was examined using two Gulf of Mexico aquatic species: (1) the mysid shrimp, *Americamysis bahia*, an aquatic invertebrate, and (2) the inland silverside, *Menidia beryllina*, a small estuarine fish. These species are standard test organisms used in a variety of EPA toxicity test methods. The static acute toxicity test methods followed, with slight modification, EPA Test Method 821-R-02-012, *Methods for Measuring the Acute Toxicity of Effluents and Receiving Waters to Freshwater and Marine Organisms* (USEPA, 2002). Reference toxicant tests were performed and reported in Phase One (Hemmer et al., 2010).

Assessment of oil-only and dispersant-oil toxicity was determined using water accommodated fractions (WAFs) of LSC or chemically-enhanced water accommodated fractions (CE-WAFs) of dispersant-LSC as described below. All tests were conducted using an established contract testing laboratory and in compliance with the Good Laboratory Practice regulations as provided in EPA 40CFR160 (USEPA, 40CFR Part 160). Two rounds of testing of the dispersant-oil mixtures were necessary to ensure that test concentrations bracketed the CE-WAF LC50 and met test condition requirements for both mysids and *Menidia*.

## 2.2. Test Organisms, Oil and Dispersants

Larval mysid shrimp were supplied from in-house cultures maintained by the contract testing laboratory using filtered natural seawater. Larval *Menidia* were purchased from Aquatic Biosystems, Inc., Fort Collins, CO, shipped overnight to the testing laboratory and held a minimum of two days prior to testing. Culture and holding conditions for both species were 25°C and 20 parts per thousand salinity. Mysids were 24-48 hours old and *Menidia* 11 or 14 days old at test initiation. All organisms for a given exposure were within 24 hours of the same age.

Non-weathered Louisiana sweet crude oil, lot # WP 681 was purchased from RT Corporation, Laramie, WY in 500 ml amber bottles and shipped directly to the testing laboratory. All dispersants were shipped directly from each manufacturer to the contract testing laboratory, logged into their test material center and maintained in accordance with GLP and Chain-ofCustody requirements.

## 2.3. Reference Toxicant Testing

Reference toxicant testing using sodium dodecyl sulfate (SDS) was conducted and reported in two phases. Phase one is presented in Hemmer et al. (2010). Phase two tests were conducted at the end of dispersant-oil testing using the same materials and methods described in Hemmer et al. (2010).

## 2.4. Preparation of Water Accommodated Fractions

LSC WAFs and dispersant/LSC CE-WAFs were prepared following the methods of the Chemical Response to Oil Spills: Ecological Effects Research Forum (CROSERF) (Singer et al., 2000) with the variable dilution modification described by Barron and Kaaihue (2003). In brief, glass aspirator bottles containing a hose bib fitted with silicon tubing and clamp at the bottom of the vessels were used to prepare the LSC WAFs. Each bottle was filled with 1 9L of seawater leaving a 20% headspace above the liquid, a stir bar added and placed on magnetic stir plate. LSC was added at 25g/L seawater using a long tube attached to a glass funnel to reduce production of air bubbles in the surface slick. The stir plate was adjusted to obtain an oil vortex of 25% of the total volume of seawater which provided a similar mixing energy in each WAF preparation. The bottles were securely covered and the solutions mixed for 18 hours then allowed to settle for 6 hours. The WAF (aqueous phase) was removed from the bottom without disturbance of the oil slick remaining on the surface. The WAF was remixed after removal and 2 liters of the WAF was used for analysis of TPH with the remaining volume available for preparation of the test solutions. Two additional LSC-only WAFs were prepared with 50 g/L and 100 g/L oil to produce higher TPH concentrations for determining the acute toxicity to *Menidia*.

The method for the preparation of each of the 8 dispersant/LSC CE-WAFs followed the LSC WAF procedure (25 g oil/L) with the addition of each dispersant at a ratio of 1:10 dispersant to oil (2.5 g/L) after the 25% oil vortex was established. Mixing and settling times followed the oil-only procedures above. Dispersant manufacturers have generally recommended application rates using dispersant to oil ratios between 1:50 to 1:10 depending on oil type and sea conditions. A ratio of 1:10 is the standard recommendation for dispersant toxicity testing because it maximizes the effect of the dispersant on oil in the CE-WAF (Barron and Kaiihue, 2003).

## 2.5. Preparation of Test Solutions

Separate oil-only WAFs were used for preparing test solutions for mysids (25 g oil/L) and *Menidia* with LSC (25, 50, 100 g oil/L). For dispersant-oil testing, each CE-WAF was divided and used to prepare solutions for both mysid and *Menidia* tests. Natural filtered seawater adjusted to 20 parts per thousand with laboratory well water was used for all static acute tests. Larval mysids and *Menidia* were treated with dilutions of LCS WAF or dispersant/oil CE-WAF plus an untreated seawater control. Six concentrations (plus control) of the oil-only WAF was tested, with the highest exposure level being 100% WAF. Each of the dispersant/oil CE-WAF tests was performed with 6 to 8 exposure concentrations (plus control) in order to bracket the median lethal concentration (Table 1).

## 2.6. Exposure Conditions

Three replicates were conducted for each exposure concentration. Test organisms were randomly assigned across exposure and control treatments with each replicate receiving 10 animals for a total of 30 animals per treatment level. One liter beakers containing 1L of test solution were maintained in 25° C temperature-controlled water baths under a photoperiod of 16 hr light:8 hr dark. All test vessels were continuously aerated (100 bubbles/min). The duration of the acute tests was 48 hrs for mysids and 96 hrs for *Menidia*. Temperature was monitored continuously using max-min thermometers; salinity and dissolved oxygen was measured once per day.

## 2.7. Analytical Chemistry

A one L sample was collected from each WAF and two replicate 1 L samples were collected from each CE-WAF for analysis of total petroleum hydrocarbons (TPH), the standard method for quantifying oil in water. Each sample was extracted with hexane, reduced to 1 mL and analyzed by GC-FID following EPA SW-846, Method 8015B –DRO. Additional analyses were performed on the CE-WAFs of Corexit 9500A and JD-2000 to provide tentative identification of single high level chemical peaks in the chromatograms. The peaks were identified as non-petroleum hydrocarbon constituents in Corexit 9500A and JD-2000 and were removed from the calculation of their respective TPH values. Final CE-WAF concentrations of TPH were determined as the average of the two replicate measured values and reported as mg TPH/L.

**Table 1. Measured concentrations, visual observations and % exposure concentrations of dispersant-oil CE-WAFs preparations used in acute toxicity tests with mysid shrimp and inland silversides**

| Dispersant -LSC CE-WAF | CE-WAF Visual Observations | Measured TPH in 100% CE-WAF (mg/L) | Definitive Test Concentrations (% CE-WAF) | |
|---|---|---|---|---|
| | | | Mysid Shrimp (*Americamysis bahia*) | Inland Silverside (*Menidia beryllina*) |
| Corexit 9500A | Very dark brown | 44.6 | 50, 25, 13, 6.3, 3.1, 1.6 | 100, 50, 25, 13, 6.3, 3.1 |
| Dispersit SPC 1000 | Cloudy beige | 400 | 3.1, 1.6, 0.8, 0.4, 0.2 0.1, 0.05 | 6.3, 3.1, 1.6, 0.8, 0.4, 0.2, 0.1, 0.05 |
| JD-2000 | Slightly cloudy with oil particulates | 6.8 | 100, 50, 25, 13, 6.3, 3.1 | 100, 50, 25, 13, 6.3, 3.1 |
| Nokomis 3-AA | Slightly cloudy beige | 87 | 3.1, 1.6, 0.8, 0.4, 0.2 0.1, 0.05 | 6.3, 3.1, 1.6, 0.8, 0.4, 0.2, 0.1, 0.05 |
| Nokomis 3-F4 | Dark cloudy brown | 1600 | 3.1, 1.6, 0.8, 0.4, 0.2 0.1, 0.05 | 3.1, 1.6, 0.8, 0.4, 0.2, 0.1, 0.05, 0.025 |
| Saf-Ron Gold | Cloudy pearlescent white | 57 (mysid) 63 (Menidia) | 50, 25, 13, 6.3, 3.1, 1.6 | 6.3, 3.1, 1.6, 0.8, 0.4, 0.2 |
| Sea Brat #4 | Slightly cloudy, brown tint | 86 | 6.3, 3.1, 1.6, 0.8, 0.4, 0.2 | 13, 6.3, 3.1, 1.6, 0.8, 0.4 |
| ZI -400 | Very cloudy brown | 1800 | 6.3, 3.1, 1.6, 0.8, 0.4, 0.2 0.1, 0.05 | 3.1, 1.6, 0.8, 0.4, 0.2, 0.1, 0.05, 0.025 |

## 2.8. Statistical Analysis

The commercially available statistical software package, CETIS® was used for the calculation of LC50 values using an automated decision tree adapted from EPA for selection of the appropriate statistical method (CETIS, 2009; USEPA, 1994). The LC50 is defined as the concentration of a substance causing mortality in 50% of test organisms for a specified time interval, in this case, 48-hours for the mysid test and 96-hours for the silverside test. The non-parametric Spearman-Karber and Trimmed Spearman-Karber methods were used to calculate the LC50 values and 95% confidence intervals. LC50 values are reported as parts per million (mg/L).

# 3. RESULTS

## 3.1. Reference Toxicant Testing with Mysid and Menidia

The LC50 values (and 95% confidence intervals) for the reference toxicant SDS wasl 8 mg/L (15-21) for mysids and 10 mg/L (8.6-12) for *Menidia*. These results were similar to reference toxicant testing in Phase I (Hemmer et al., 2010), indicating similar sensitivity of test organisms over the course of the experiments.

## 3.2. Oil-only Acute Tests with Mysid and Menidia

### *3.2.1. Test Acceptability*

Control performance met all criteria for an acceptable exposure in each test (≥90% survival) for both mysid and *Menidia* exposures. All water quality parameters in all treatments were within ranges specified in the protocol for each species.

### *3.2.2. Measured TPH and toxicity results*

The measured TPH concentration in the LSC WAF used for the acute mysid test was 4.4 mg/L, resulting in a calculated LC50 value of 2.7 mg/L and corresponding 95% confidence interval of 2.5-3.0 mg/L. The mean measured TPH concentrations in LSC WAF used for the acute tests with *Menidia* were 2.9, 5.4, and 5.1 mg/L for oil-only WAFs prepared with 25, 50, and 100 g oil/L, respectively. A definitive LC50 for *Menidia* was determined at 100 g oil/L to be 3.5 mg TPH/L with a corresponding confidence interval of 3.4-3.7 mg/L.

**Table 2. EPA five-step scale of toxicity categories used to classify chemicals based on their acute toxicity (USEPA, 2010b)**

| LC50 (ppm) | Toxicity Classification |
|---|---|
| > 100 | Practically Nontoxic |
| >10 to 100 | Slightly Toxic |
| >1 to 10 | Moderately Toxic |
| 0.1 to 1.0 | Highly Toxic |
| <0.1 | Very Highly Toxic |

EPA uses a five-step scale of toxicity categories to classify pesticides based on their acute toxicity to aquatic organisms which is provided in Table 2 (USEPA, 20 10b). Using this toxicity classification, LSC would be classified as moderately toxic to both mysids and *Menidia.*

## 3.3. Measured TPH in Dispersant-Oil CE- WAF

TPH in 100% CE-WAF ranged from 6.8 mg/L for JD-2000 to 1800 mg/L for ZI-400 (Table 1). The CE-WAFs for Corexit 9500A and JD-2000 each contained a single large non- petroleum hydrocarbon chemical peak. The chromatogram peak areas associated with these compounds were removed from the calculation of TPH prior to determination of the LC50 values and 95% confidence intervals for Corexit 9500A and JD-2000.

## 3.4. Dispersant-Oil Toxicity Tests - Mysids

### *3.4.1. Acceptability of Initial Tests*

Control performance met all criteria for an acceptable exposure (≥90% survival) in all initial tests. A definitive LC50 was determined for the initial test with Saf-Ron Gold CE-WAF and all water quality parameters were within ranges specified in the protocol for all treatments. Initial tests with CE-WAFs of Dispersit SPC 1000, Nokomis 3-AA, Nokomis 3-F4 and ZI-400 had LC50s less than the lowest concentration tested which required a second round of testing. A second round of testing was also required for Corexit 9500A, JD-2000 and Sea Brat #4 because of temperature or dissolved oxygen deviations from protocols.

### *3.4.2. Acceptability of Second Round Tests*

Control performance met all criteria for an acceptable exposure (≥ 90% survival) in each of the CE-WAF tests conducted in round two. All water quality parameters were within ranges specified in the test protocol for mysids and definitive LC50 values were determined for each of the seven dispersant-oil CE-WAFs tested.

### *3.4.3. Mysid Definitive Test Results*

The LC50 values for dispersant-oil acute toxicity tests with mysids ranged from 0.39 to 9.7 mg TPH/L for Nokomis 3-AA and ZI-400, respectively (Table 3). Using the EPA toxicity classification, oil-dispersant mixtures using Nokomis 3-AA would be considered highly toxic, whereas Corexit 9500A, Dispersit SPC 1000, JD-2000, Nokomis 3-F4, Saf-Ron Gold, Sea Brat #4 and ZI-400 would be classified as moderately toxic to mysid shrimp.

## 3.5. Dispersant-Oil Toxicity Tests – Menidia

### *3.5.1. Acceptability of Initial Tests*

Control performance met all criteria for an acceptable exposure (≥90% survival) in all initial tests. Initial tests with CE-WAFs of Dispersit SPC 1000, Nokomis 3-AA, Nokomis 3-F4 Saf-Ron Gold and ZI-400 had LC50s less than the lowest concentration tested, which required a second round of tests to be conducted. Second round testing was also required for Corexit 9500A, JD-2000 and Sea Brat #4 due to temperature or dissolved oxygen deviations from protocols.

### *3.5.2. Acceptability of Second Round Tests*

Control performance met all criteria for an acceptable exposure (≥ 90% survival) in each of the CE-WAF tests conducted in round two. All water quality parameters were within ranges specified in the test protocol for *Menidia*. In the second round of acute tests, LC50 values and 95% confidence intervals were successfully determined for all eight dispersant-oil CE-WAFs.

### *3.5.3. Menidia Definitive Test Results*

The LC50 values for dispersant-oil acute toxicity tests with *Menidia* ranged from 0.64 for Dispersit SPC 1000 to 13.1 mg TPH/L for ZI-400 (Table 3). Using the EPA toxicity classification, oil-dispersant mixtures using Dispersit SPC 1000 or Nokomis 3-AA would be considered highly toxic, whereas Corexit 9500A, JD-2000, Nokomis 3-F4, Saf-Ron Gold and Sea Brat #4 would be classified as moderately toxic and ZI-400 as slightly toxic to inland silversides.

**Table 3. Results of mysid 48-hr and *Menidia* 96-hr acute toxicity tests with eight dispersant-oil mixtures (CE-WAFs) and oil-only (WAF), LC50 values, 95% confidence intervals [in brackets] and their toxicity classification derived in the present study. Results of LC50s and 95% confidence intervals [in brackets] from Phase One dispersant-only acute tests are supplied for comparison**

| Dispersant CE-WAF | Dispersant[1] LC50 (ppm)[2] [95% CI] | Dispersant Toxicity Category | CE-WAF LC50 (ppm)[3] [95% CI] | CE-WAF Toxicity Category | Dispersant1 LC50 (ppm)[2] [95% CI] | Dispersant Toxicity Category | CE-WAF LC50 (ppm)[3] [95% CI] | CE-WAF Toxicity Category |
|---|---|---|---|---|---|---|---|---|
| Corexit 9500A | 42 [38-47] | Slightly Toxic | 5.4 [4.9-6.7] | Moderately Toxic | 130 [122-138] | Practically Nontoxic | 7.6 [6.2-8.5.] | Moderately Toxic |
| Dispersit SPC 1000 | 12 [10-14] | Slightly Toxic | 2.1 [1.9-2.2] | Moderately Toxic | 2.9 [2.5-3.2] | Moderately Toxic | 0.64 [0.56-0.72] | Highly Toxic |
| JD-2000 | 788 [627-946] | Practically Nontoxic | 1.4 [1.2-1.6] | Moderately Toxic | >5600 [NA] | Practically Nontoxic | 4.6 [4.3-4.9] | Moderately Toxic |
| Nokomis 3-AA | 30 [27-34] | Slightly Toxic | 0.39 [0.35-0.44] | Highly Toxic | 19 [17-21] | Slightly Toxic | 0.96 [0.96-1.0] | Highly Toxic |
| Nokomis 3-4F | 42 [38-47] | Slightly Toxic | 7.4 [6.6-8.3] | Moderately Toxic | 19 [16-21] | Slightly Toxic | 9.6 [8.2-11.2] | Moderately Toxic |
| Saf-Ron Gold | 118 [104-133] | Practically Nontoxic | 2.9 [2.6-3.1] | Moderately Toxic | 44 [41-47] | Slightly Toxic | 1.6 [1.5-1.8] | Moderately Toxic |
| Sea Brat #4 | 65 [57-74] | Slightly Toxic | 1.4 [1.2-1.6] | Moderately Toxic | 55 [49-62] | Slightly Toxic | 3.4 [3.0-3.7] | Moderately Toxic |
| ZI-400 | 55 [50-61] | Slightly Toxic | 9.7 8.1-11.7 | Moderately Toxic | 21 [18-23] | Slightly Toxic | 13.1 [11.2-15.3] | Slightly Toxic |
| **Oil-only WAF** | | | **Oil LC50 (ppm) [95% CI]** | **Oil Toxicity Category** | | | **Oil LC50 (ppm)** | **Oil Toxicity Category** |
| Louisiana Sweet Crude | | | 2.7 [2.5-3.0] | Moderately Toxic | | | 3.5 [3.4-3.7] | Moderately Toxic |

[1] From Hemmer et al., 2010; [2] ppm = μL/L; [3] ppm = mg TPH/L.

# CONCLUSION

The present study provided an independent, consistent, and quantitative assessment of acute toxicities of Louisiana sweet crude (LSC) oil and eight dispersant-LSC mixtures to two aquatic species inhabiting Gulf of Mexico waters. Toxicity was determined as the LC50 (mg TPH/L or ppm) derived from standard short term acute tests using standard test species, specifically the Gulf mysid, *Americamysis bahia*, and the inland silverside, *Menidia beryllina*. LSC oil alone was found to be moderately toxic to both mysids (LC50 = 2.7 mg TPH/L) and *Menidia* (LC50 = 3.5 mg TPH/L). In the dispersant-LSC acute tests, the toxicity values for mysids and *Menidia* ranged over two orders of magnitude (0.39 to 13.1 mg TPH/L). For mysid shrimp, Nokomis 3-AA was the most toxic and ZI-4000 the least toxic. For *Menidia*, Dispersit SPC 1000 and Nokomis 3-AA were the most toxic and ZI-4000 the least toxic of the dispersant-oil mixtures. The other five dispersant/LSC mixtures had generally similar toxicity to mysids and *Menidia*, with LC50 values ranging from 1.4 to 9.7 mg TPH/L.

Overall, the dispersants/LSC mixtures were classified as being highly toxic to moderately toxic depending on the test species and dispersant. The ZI-400/LSC mixture was the exception and would be considered only slightly toxic to *Menidia*. Corexit 9500A, the dispersant that has been applied offshore at the surface and in the deep ocean, falls into the moderately toxic category for both species. For all eight dispersants in both test species, the dispersants alone were less toxic than the dispersant-oil mixture. For mysids and *Menidia*, LSC had greater toxicity (lower LC50s) than the eight dispersants when tested alone. LSC alone had similar toxicity to mysids as the dispersant-oil mixtures, with exception of greater toxicity of the Nokomis 3-AA/LSC mixture. LSC alone had similar toxicity to *Menidia* as the dispersant-oil mixtures, with exception of greater toxicity of the Dispersit SPC 1000/LSC and Nokomis 3-AA/LSC mixtures.

Short-term acute toxicity tests using consistent methodologies and test organisms provide important and fundamental information on oil spill dispersants and other toxicants. The comparative toxicity analysis of dispersants, sweet crude oil and dispersant-sweet crude oil mixtures on standard aquatic test species provide an improved understanding of potential toxicological effects associated with this oil spill and help inform future decision-making.

## REFERENCES


Barron, M. G. & Kaaihue, L. (2003). Critical evaluation of CROSERF test methods for oil dispersant toxicity testing under subarctic conditions. *Marine Poll. Bull., 46*, 1191-1199

CETIS. (2009). *Comprehensive Environmental Toxicity Information System: Users Manual.* Tidepool Scientific Software, McKinleyville, CA.

Hemmer, M. J., Barron, M. G. & Greene, R. M. (2010). *Comparative Toxicity of Eight Oil Dispersant Products on Two Gulf of Mexico Aquatic Test Species.* (http://www.epa.gov/bpspill/reports/ Comparative Tox Test.Final.6. 30.1 0.pdf).

Judson, R. S., Martin, M. T., Reif, D. M., Houck, K. A., Knudsen, T. B., Rotroff, D. M., Xia, M., Sakamuru, S., Huang, R., Shinn, P., Austin, C. P., Kavlock, R. J. & Dix, D. J. (2010). Analysis of Eight Oil Spill Dispersants Using Rapid, In Vitro Tests for Endocrine and Other Biological Activity. *Environ. Sci. Technol., 44*, 597 1-5978. (http://pubs.acs.org/doi/abs/10.1021/es102150z.

Singer, M. M., Aurand, D., Bragins, G. E., Clark, J. R., Coelho, G. M., Sowby, M. L. & Tjeerdema, R. S. (2000). Standardization of preparation and quantification of water-accommodated fractions of petroleum for toxicity testing. *Mar Pollut Bull, 40*, 1007-1016.

USEPA. 40 CFR, Part 160. *Federal Insecticide, Fungicide, and Rodenticide Act.* Good Laboratory Practices Standards; Final Rule. Office of the Federal Register, National Archives and Records Administration. U.S. Government Printing Office, Washington, D.C.

USEPA. (1994). *Short-Term Methods for Estimating the Chronic Toxicity of Effluents and Receiving Waters to Marine and Estuarine organisms,* 2nd ed. Environmental Monitoring Systems Laboratory, Office of Research and Development, U.S. Environmental Protection Agency, Cincinnati, OH. EPA/600/4-51/003.

USEPA. (2002). *Methods for Measuring the Acute Toxicity of Effluent and Receiving Waters to Freshwater and Marine Organisms.* 5th Edition, October 2002. EPA-821-R-02-012. U.S. Environmental Protection Agency, Washington, D.C.

USEPA. (2010a). http://www.epa.gov/emergencies/content/ncp/product schedule.htm *USEPA.* 2010b. http://www.epa.gov/oppefed1/ecorisk_ders/toera_analysis_eco.htm#Ecotox. Accessed June 23, 2010.

In: Dispersants in Oil Spills: Issues and Research ISBN: 978-1-61209-241-6
Editors: Edward K. Holm 

*Chapter 4*

# COMPARATIVE TOXICITY OF EIGHT OIL DISPERSANT PRODUCTS ON TWO GULF OF MEXICO AQUATIC TEST SPECIES

***United States Environmental Protection Agency***

## 1. INTRODUCTION

Large quantities of Louisiana sweet crude oil have been released into the Gulf of Mexico since the explosion of the Deepwater Horizon oil exploration platform on April 20, 2010. As part of the integrated response effort to mitigate the impact of the oil in the environment, the decision was made to use dispersants listed on the U.S. Environmental Protection Agency's (EPA) National Contingency Plan (NCP) Product Schedule (EPA 2010a). Dispersants are being applied offshore on the surface as well as underwater at the source of the leak. The EPA conducted independent studies to assess the relative acute toxicity of eight dispersants on the NCP Product Schedule.

This report summarizes results of the first phase of testing obtained from acute toxicity tests conducted with eight oil dispersants using two Gulf of Mexico aquatic species: (1) the mysid shrimp, *Americamysis bahia*, an aquatic invertebrate, and (2) the inland silverside, *Menidia beryllina*, a small estuarine fish. These species are standard test organisms used in a variety of EPA toxicity test methods. The eight dispersants tested were Corexit 9500A, Dispersit SPC 1000, JD-2000, Nokomis 3-AA, Nokomis 3-F4, Saf-Ron Gold,

Sea Brat #4 and ZI-400. The tests were conducted using an established contract testing laboratory and in compliance with the Good Laboratory Practice regulations as provided in EPA 40CFR160 (USEPA, 40CFR Part 160). The approach described herein utilized consistent test methodologies within a single laboratory which provided a means to assess acute toxicity estimates across dispersants and independently evaluate the NCP Product schedule toxicity information. The next phases of this study will examine the acute toxicity of Louisiana sweet crude oil and dispersant-sweet crude oil mixtures on mysids and *Menidia* – the results will be reported separately at a later date.

## 2. Test Methods

The acute toxicity test methods followed, with slight modification, the requirements specified in U.S. Environmental Protection Agency's 62 FR 15576, Appendix C of Part 300 – *Swirling Flask Dispersant Effectiveness Test, Revised Standard Dispersant Toxicity Test and Bioremediation Agent Effectiveness Test* (USEPA, 1997) and the EPA Test Method 821 -R-02- 012, *Methods for Measuring the Acute Toxicity of Effluents and Receiving Waters to Freshwater and Marine Organisms* (USEPA, 2002). Specific modifications are shown in Appendix A.

The exposure concentration range for each dispersant was chosen to bracket the estimated median lethal concentration (LC50) values reported in the NCP Product Schedule. The LC50 is defined as the concentration of a substance causing mortality in 50% of test organisms for a specified time interval, in this case, 48-hours for the mysid test and 96-hours for the silverside test. The commercially available statistical software package, CETIS® was used for the calculation of LC50 values using an automated decision tree adapted from EPA for selection of the appropriate statistical method (CETIS, 2009; USEPA, 1994). Point estimate procedures used to calculate LC50 values included linear regression methods, the non-parametric SpearmanKarber method and the binomial method. A qualitative comparison was made between LC50 values for the eight dispersants tested as well as with those available in the NCP Product Schedule. Note that the reproducibility of static acute tests among laboratories using the same species/toxicant combination has been reported to generally fall within a factor of 3.5 among laboratories when using nominal concentrations (unmeasured treatment concentrations) for both freshwater and marine species (USEPA, 1981). Given the use of whole organisms in these tests, some variation in response attributable to differences in parameters such as culture and acclimation conditions, stock populations or variable water quality is expected and acceptable.

# 3. Results - Mysid Toxicity Tests

## 3.1. Mysid Testing Schedule

Following the first round of eight acute toxicity tests, dispersant LC50s were greater than the highest concentration tested for four of the eight dispersants. Definitive acute toxicity tests were repeated using higher test concentrations for JD-2000, Saf-Ron Gold, Sea Brat #4 and ZI400.

## 3.2. Mysid Test Acceptability

Control performance (without dispersant) met all criteria for an acceptable exposure in each test (≥90% survival). All water quality parameters were within ranges specified in the protocol with the exception of dissolved oxygen for the high test concentration (56 ppm) in the Nokomis 3-AA exposure at 24 hours, which was measured at 56% of saturation. As dissolved oxygen levels were >60% at other time points in the test and the toxicity was clearly dose related, the departure observed in the 56 ppm concentration at 24 hours was not considered to have had a negative impact on the exposure with Nokomis 3-AA.

## 3.3. Mysid Toxicity Results

In the first series of acute toxicity tests, LC50 values and 95% confidence intervals were successfully determined for Corexit 9500A, Dispersit SPC 1000, Nokomis 3-AA and Nokomis 3-F4 and in the second series of acute tests, LC50s were calculated for JD-2000, Saf-Ron Gold, Sea Brat #4 and ZI400. Test results are summarized in Table 1.

**Table 1. Results of mysid 48-hr static acute toxicity tests with eight dispersants. LC50 values (ppm), 95% confidence intervals [in brackets] and the toxicity classification of dispersant LC50s derived in the present study. NCP Product Schedule listing of dispersant LC50s and 95% confidence intervals [in brackets] shown in right column for comparison**

| Dispersant | This Study LC50 (ppm) [95% CI] | Toxicity Category[1] | NCP Product Schedule LC50 (ppm) [95% CI][d] |
|---|---|---|---|
| Dispersit SPC 1000 | 12 [10-14][a] | Slightly Toxic | 16.6 [14.1-19.6] |
| Nokomis 3-AA | 30 [27-34][b] | Slightly Toxic | 20.2 [17.4-22.8] |
| Corexit 9500A | 42 [38-47][c] | Slightly Toxic | 32.2 [26.5-39.2] |
| Nokomis 3-F4 | 42 [38-47][c] | Slightly Toxic | 32.2 [28.4-36.5] |
| ZI -400 | 55 [50-61][b] | Slightly Toxic | 21.0 [17.9-24.5] |
| Sea Brat #4 | 65 [57-74][a] | Slightly Toxic | 14.0 [+10.4] |
| Saf-Ron Gold | 118 [104-133][b] | Practically Non-Toxic | 63.0[e] [52.9-75.1] |
| JD-2000 | 788 [627-946][a] | Practically Non-Toxic | 90.5[e] [76.1-108] |

[1] Toxicity classification per USEPA 2010 applied to results of present study
[a] Estimated by linear regression method
[b] Estimated by Spearman-Karber method
[c] Estimated by binomial method
[d] Values as reported in NCP Product Schedule documentation by manufacturer eClassified as slightly toxic according to values provided in NCP Product Schedule

The LC50 values for dispersant acute tests with mysids ranged from 12 ppm for Dispersit SPC1000 to 788 ppm for JD-200 (Table 1). EPA uses a five-step scale of toxicity categories to classify pesticides based on their acute toxicity to aquatic organisms: LC50 values of >100 ppm are considered practically nontoxic; >10 to 100 ppm as slightly toxic; > 1 to 10 ppm as moderately toxic; LC50s of 0.1 to 1 ppm as highly toxic and LC50s < 0.1 ppm as very highly toxic (USEPA, 20 10b). Using this toxicity classification, Corexit 9500A, Dispersit SPC 1000, Nokomis-3AA, Nokomis 3-F4, Sea Brat #4 and ZI-400 would be classified as slightly toxic whereas JD-2000 and Saf-Ron Gold would be classified as practically non-toxic to mysids (Table 1).

Based on comparison of LC50 values and 95% confidence intervals across the eight dispersants tested in the present study, the rank order toxicity (most to least toxic) of the dispersants to mysids was: (1) Dispersit SPC 1000, (2) Nokomis 3-AA, (3) Nokomis 3-F4, Corexit 9500A, (4) ZI-400, Sea Brat #4, (5) Saf-Ron Gold, and (6) JD-2000.

Factor ratios were used to compare LC50s derived for the same species/dispersant combination from different laboratories. The factor ratios between LC50 values determined in this study and NCP reported LC50 values were calculated as a ratio by dividing the higher of the two LC50 values by the lower LC50 value for each of the eight dispersants, respectively (Table 1). As an example, using information from Table 1, the factor ratio for Corexit 9500A was determined as 42/32.2 = 1.3. The factor ratios calculated for Corexit 9500A, Dispersit SPC1000, Nokomis 3-AA, Nokomis 3-F4, Saf-Ron Gold and ZI-400 were less than or equal to 2.6 which was considered within normal inter-laboratory variability (USEPA, 1981). Results for JD-2000 and Sea Brat #4 showed lower toxicities (i.e., higher LC50s) with factor ratios of 8.7 and 4.6, respectively, compared to their reported NCP LC50 values.

### 3.4. Mysid Reference Toxicant Test

A 48-hr acute toxicity test was conducted with the standard reference toxicant, sodium dodecyl sulfate (SD S), to evaluate the relative sensitivity of the mysids used in the series of dispersant toxicity tests. The mysids tested with SDS were from the same population and age range used for dispersant testing. The 48-hr LC50 and 95% confidence interval calculated for SDS was 23 ppm [19-26 ppm] which was consistent with the reported NCP LC50 values for SDS.

## 4. RESULTS - *MENIDIA* TOXICITY TESTS

### 4.1. *Menidia* Testing Schedule

Following the first round of acute toxicity tests, dispersant LC50s were determined to be greater than the highest concentration tested for two of the eight dispersants. Definitive acute toxicity tests were repeated using higher test concentrations for Corexit 9500A and JD-2000.

## 4.2. *Menidia* Test Acceptability

Control performance met all criteria for an acceptable exposure in each of the eight dispersant tests conducted (≥ 90%). All water quality parameters were within ranges specified in the test protocol for *Menidia beryllina.*

## 4.3. *Menidia* Toxicity Results

In the first series of acute tests, LC50 values and 95% confidence intervals were successfully determined for Dispersit SPC 1000, Nokomis 3-AA, Nokomis 3-F4, Saf-Ron Gold, Sea Brat #4 and ZI-400. In the second series of repeat acute tests, an LC50 was calculated for Corexit 9500A but not for the dispersant JD-2000. These data are summarized in Table 2. In the case of JD-2000, 20% mortality was observed in the highest concentration tested of 5,600 ppm, followed by no mortality observed in the next two highest exposure concentrations which indicated an LC50 > 5,500 ppm. At the highest concentration, solid material was observed at the bottom of the replicate test vessels suggesting saturation of the dispersant may have been achieved.

The LC50 values for dispersant acute toxicity tests with *Menidia* ranged from 2.9 ppm for Dispersit SPC1000 to 130 ppm for Corexit 9500A; the LC50 for JD 2000 exceeded the highest test concentration of 5,600 ppm. Using the EPA toxicity classification, Dispersit SPC 1000 would be considered moderately toxic whereas Nokomis-3AA, Nokomis 3-F4, SafRon Gold, Sea Brat #4 and ZI-400 would be classified as slightly toxic, and Corexit 9500A and JD-2000 as practically non-toxic to inland silversides.

Based on comparison of LC50 values and 95% confidence intervals, the rank order toxicity (most to least toxic) of the dispersants to *Menidia* were: (1) Disersit SPC 1000, (2) Nokomis 3-F4, Nokomis 3-AA, ZI-400, (3) Saf-Ron Gold, (4) Sea Brat #4, (5) Corexit 9500A, and (6) JD-2000.

The factor ratios calculated for Dispersit SPC 1000, Nokomis 3-AA, Nokomis 3-F4, SafRon Gold, Sea Brat #4 and ZI-400 were less than or equal to 1.83 which was considered within normal inter-laboratory variability. The factor ratios of 5.2 and 13.8 for Corexit 9500A and JD- 2000 indicate that the LC50 values reported for Corexit 9500A and JD-200 in the NCP Product Schedules would be considered different (i.e., lower) from the LC50 values determined in the present study.

**Table 2. Results of *Menidia* 96-hr static acute toxicity tests with eight dispersants. LC50 values (ppm), 95% confidence intervals [in brackets] and the toxicity classification of dispersant LC50s derived in the present study. NCP Product Schedule listing of dispersant LC50s and 95% confidence intervals [in brackets] shown in right column for comparison**

| Dispersant | This Study LC50 (ppm) [95% CI] | Toxicity Category[1] | NCP Product Schedule LC50 (ppm) [95% CI][d] |
|---|---|---|---|
| Dispersit SPC 1000 | 2.9 [2.5-3.2][b] | Moderately Toxic | 3.5 [3.1-4.0] |
| Nokomis 3-F4 | 19 [16-21][b] | Slightly Toxic | 29.8 [24.0-35.4] |
| Nokomis 3-AA | 19 [17-21][b] | Slightly Toxic | 34.2 [29.2-37.95] |
| ZI -400 | 21 [18-23][b] | Slightly Toxic | 31.8 [28.7-35.1] |
| Saf-Ron Gold | 44 [41-47][b] | Slightly Toxic | 29.4 [25.2-34.3] |
| Sea Brat #4 | 55 [49-62][b] | Slightly Toxic | 30.0 [+16.2] |
| Corexit 9500A | 130 [122-138][b] | Practically Non-Toxic | 25.2[c] [13.6-46.6] |
| JD-2000 | >5,600 | Practically Non-Toxic | 407 [330-501] |

[1] Toxicity classification per USEPA 2010 applied to results of present study
[b] Estimated by Spearman-Karber method
[d] Values as reported in NCP Product Schedule documentation by manufacturer
[c] Classified as slightly toxic according to values provided in NCP Product Schedule

Possible explanations for the 13.8 fold difference between the reported NCP LC50 for JD-2000 and the highest exposure concentration tested in the present study may be attributable to batch-to-batch variability in the manufacturing process, instability of the stored product over time, or a change in the product formulation.

## 4.4. *Menidia* Reference Toxicant Test

A 96-hr acute toxicity test was conducted with the reference toxicant SDS to evaluate the relative sensitivity of the *Menidia* used in the series of dispersant toxicity tests. The *Menidia* tested with SDS were from the same population and age range used for dispersant testing. The 96-hr LC50 and 95% confidence interval calculated for SDS was 9.5 ppm [8.7-10 ppm] which was consistent with the reported NCP LC50 values for SDS. It should be noted that during the last 24 hours of the test, the temperature dropped to 22°C, which was 2 degrees below the acceptable criteria and thus invalidated the test. However, there was no difference in mortality counts between the 72-hour and the 96-hour observations suggesting the temperature change had no negative impact on the test or the final calculated LC50.

## CONCLUSION

The present study provided an independent, quantitative assessment of acute toxicities of eight dispersants to two aquatic species inhabiting Gulf of Mexico waters. Toxicity was determined as the LC50 derived from standard short term acute tests using standard test species, specifically the Gulf mysid, *Americamysis bahia*, and the inland silverside, *Menidia beryllina*. In general, the toxicity values (i.e., LC50s) for mysids ranged over nearly two orders of magnitude and for *Menidia* over three orders of magnitude. Given the expected range of inter-laboratory variability, the results of the present study were consistent with test results reported in the NCP Product Schedule, with the exception of two dispersants for each test species which yielded higher LC50s (i.e., lower toxicity) than reported in the NCP. The rank order toxicity of the eight dispersants was generally similar to the information provided in the NCP Product Schedule. For both test species, Dispersit SPC 1000 was the most toxic and JD-2000 the least toxic. The other six dispersants varied in relative toxicity to mysids and *Menidia*, with LC50 values ranging from 20 to 130 ppm. Overall, the dispersants were classified as being slightly toxic to practically nontoxic to both test species, with the exception that Dispersit SPC 1000 would be considered moderately toxic to *Menidia*. Corexit 9500A, the dispersant currently applied offshore at the surface and underwater, falls into the slightly toxic category for mysids and the practically non toxic category for *Menidia*.

Short-term acute toxicity tests using consistent methodologies and test organisms provide important and fundamental information on oil spill dispersants and other toxicants. The next phase of testing will examine the acute toxicity of Louisiana sweet crude oil and dispersant- sweet crude oil mixtures on mysids and *Menidia*. The comparative toxicity analysis of dispersants, sweet crude oil and dispersant-sweet crude oil mixtures on standard aquatic test species will provide improved understanding of potential toxicological effects associated with this oil spill.

# APPENDIX A

| Test parameter | Specified in SubPart J Appendix C (USEPA 1997) | Method used in present study, and specified in USEPA 2002 |
|---|---|---|
| Photoperiod and light intensity | *24 hr light<br>*higher intensity light | *16 hr light/8 hr dark<br>*Moderate intensity light |
| Glassware cleaning | *Hexane immersion | *Acetone rinse |
| Reference toxicant test | *Two species simultaneously | *Staggered tests |
| Rangefinder tests | *Prior to definitive test | *Use NCP data to define test concentrations |
| Mysid age | *5-7 day old larvae | *1 to 6 day old; all within 24 hr same age |
| Toxicant stock solution preparation for mysid test | *Blender 10,000 rpm<br>*gas tight syringes | *Top stirring at 70% vortex<br>*graduated glass pipettes |
| Mysid test solution mixing | *no specification | *short term gentle mixing following stock addition |
| Mysid additions to test chambers | *no specification | *impartial, two at a time |
| *Menidia* age | *7 day old larvae | *9-14 day old, all within 24 hr same age |
| *Menidia* test solution mixing | *test jars on shaker platform | *same procedure as for mysids |
| Dilution Water | *Natural Seawater Prefered | *Salinity adjusted, 20 μm filtered natural seawater |

# REFERENCES

CETIS. (2009). *Comprehensive Environmental Toxicity Information System: Users Manual.* Tidepool Scientific Software, McKinleyville, CA.

USEPA. 40 CFR, Part 160. *Federal Insecticide, Fungicide, and Rodenticide Act.* Good Laboratory Practices Standards; Final Rule. Office of the Federal Register, National Archives and Records Administration. U.S. Government Printing Office, Washington, D.C.

USEPA. (1981). *Results: Interlaboratory Comparison-Acute Toxicity Tests Using Estuarine Animals.* February 1981. EPA-600/4-81-003. U.S. Environmental Protection Agency, Gulf Breeze, Florida.

USEPA. (1994). *Short-Term Methods for Estimating the Chronic Toxicity of Effluents and Receiving Waters to Marine and Estuarine organisms,* 2nd ed. Environmental Monitoring Systems Laboratory, Office of Research and Development, U.S. Environmental Protection Agency, Cincinnati, OH. EPA/600/4-51/003.

USEPA. (1997). *Swirling Flask Dispersant Effectiveness Test, Revised Standard Dispersant Toxicity Test, and Bioremediation Agent Effectiveness Test.* Part 300, Appendix C. EPA 62 FR 15576. U.S. Environmental Protection Agency, Washington, D.C. April 1, 1997.

USEPA. (2002). *Methods for Measuring the Acute Toxicity of Effluent and Receiving Waters to Freshwater and Marine Organisms.* 5th Edition, October 2002. EPA-821-R-02-012. U.S. Environmental Protection Agency, Washington, D.C.

USEPA. (2010a). http://www.epa.gov/emergencies/content/ncp/product schedule.htm

USEPA. (2010b). http://www.epa.gov/oppefed1/ecorisk ders/toera analysis eco.htm#Ecotox. Accessed June 23, 2010.

In: Dispersants in Oil Spills: Issues and Research ISBN: 978-1-61209-241-6
Editors: Edward K. Holm 

*Chapter 5*

# Dispersant Toxicity Testing: Phase II Questions and Answers

*Edward K. Holm*

**Q 1. What tests were conducted in Phase II of dispersant toxicity testing?**

EPA conducted acute toxicity tests of 1).multiple concentrations of Louisiana Sweet Crude Oil alone, and 2.) combinations of Louisiana Sweet Crude Oil with each of the eight dispersants. The organisms tested were two Gulf of Mexico aquatic species: (1) the mysid shrimp, *Americamysis bahia*, an aquatic invertebrate, and (2) the inland silverside, *Men idia beryllina*, a small estuarine fish. These species are standard test organisms used in toxicity test for a variety of pollutants. The tests were conducted on mixtures of Louisiana Sweet Crude Oil and eight dispersant products found on the National Contingency Plan Product Schedule – Dispersit SPC 1000, Nokomis 3-F4, Nokomis 3-AA, ZI-400, SAFRON Gold, Sea Brat #4, Corexit 9500 A and JD 2000.

**Q 2. What did the test results show?**

The results indicate that the eight dispersants tested are similar to one another based on standard toxicity tests on sensitive aquatic organisms found in the Gulf. These results confirm that the dispersant used in response to the oil spill in the Gulf, Corexit 9500A, is no more or less toxic than the other available alternatives.

**Q3. Is there a difference in toxicity between oil alone and mixtures of oil and dispersants?**

For all eight dispersants in both test species, the dispersants alone were less toxic than the dispersant-oil mixture. Oil alone was found to be more toxic to mysid shrimp than the eight dispersants when tested alone. Oil alone had similar toxicity to mysid shrimp as the dispersant-oil mixtures, with exception of the mixture of Nokomis 3-AA and oil, which was found to be more toxic. The oil results for small fish were inconclusive. EPA will perform additional testing of the toxicity of oil to small fish.

**Q 4. Why were the oil results for the small fish inconclusive?**

For the highest concentration of oil tested, only 7 % of the the inland silverside, the small estuarine fish, died. To estimate the LC50 – the goal of this standard toxicity test – 50% mortality is needed. The test was conducted over a range of five concentrations and at the highest concentration only 7% mortality was achieved. The test will be repeated using a series of oil in water concentrations with results that encompass 50% mortality of the test organisms.

**Q 5. What tests did EPA use to assess acute toxicity to shrimp and small fish?**

**A:** Acute toxicity tests are used to determine lethal concentrations of the test chemicals. The acute toxicity to shrimp and fish was determined using a standardized 48-hour mysid shrimp and a 96-hour small fish test to evaluate the potential toxicity of the dispersant. Established testing procedures are specified for both the mysid shrimp and small fish. Both species live in the bays and estuaries of the northern Gulf of Mexico and are commonly used in toxicity tests. These two tests are required by EPA to list a dispersant on the **National Contingency Plan Product Schedule**. The test protocol exposes mysid shrimp or small fish to a range of dispersant concentrations and dispersant-oil mixtures or oil alone separately in the laboratory. Toxicity is determined by comparing the survival of the mysid shrimp or small fish exposed to the dispersants, dispersant-oil mixtures, or oil alone to survival of these organisms kept in clean, untreated seawater. The aquatic organisms used as test species are small mysid shrimp, *Americamysis bahia*, and a small fish, *Men idia beryllina.* Survival of the organisms exposed to multiple concentrations of the dispersants, dispersant-oil mixtures, or oil alone is determined for each species. The concentration lethal to 50 percent of the test organisms is calculated and compared between dispersants and between the toxicities of the oil- dispersant mixtures to determine the most and least toxic chemicals and combinations.

**Q6. Why did EPA only test eight out of the total 14 dispersants on the National Contingency Plan product schedule?**

EPA chose eight dispersants (Dispersit SPC 1000; Nokomis 3- F4; Nokomis 3-AA; ZI-400; SAFRONGOLD; Sea Brat #4; Corexit 9500 A; JD 2000) from the dispersants listed on the National Contingency Plan Product Schedule based on three criteria: 1) lower toxicity of the dispersant or of the dispersant when mixed with oil; 2) availability of sufficient quantities to respond to the Gulf spill; and 3) immediate availability of samples for testing.

## PHASE I- DISPERSANT TESTING- QUESTIONS AND ANSWERS

**Q7. What toxicity tests were conducted to determine the least toxic dispersants?**

**A:** EPA conducted several toxicity tests to provide independent scientific information about these eight dispersants. Three types of testing results on the dispersants alone are available from Phase I of the testing:

1) Potential endocrine activity: Some of the dispersants include chemicals called nonylphenol ethoxylates (NPE). NPE breaks down in the environment to nonylphenol (NP) which is a substance that could potentially cause endocrine disruption.
2) Degree each is toxic to living cells – cytotoxicity: EPA used *in vitro* assays to test the degree to which these eight dispersants are toxic to various types of mammalian cells and the potential for each dispersant to exhibit endocrine activity.
3) Acute toxicity to shrimp and small fish -Acute toxicity tests are used to determine lethal concentrations of the test chemicals.

The companies who manufacture the different types of oil spill dispersants already tested both the toxicity and the effectiveness of each of these dispersants and submitted results to EPA for listing their product on the National Contingency Plan Product Schedule. Although these industry-submitted test results provide guidance, the tests were conducted on the dispersants by different laboratories and on the dispersants mixed with No. 2 fuel oil which is not the type of oil in the Gulf. EPA wanted to conduct its own toxicity tests in one laboratory under EPA oversight for better comparative analysis and to test the dispersants mixed with the oil from the Gulf.

**Q8. What tests did EPA use to assess potential endocrine activity?**

**A:** Some of the dispersants include chemicals called nonylphenol ethoxylates (NPE). NPE breaks down in the environment to nonylphenol (NP) which is a substance that could potentially cause endocrine disruption. Endocrine disruption can lead to defects in fetal development or can impair reproductive health in humans and aquatic species. The degree to which the eight types of oil spill dispersants are toxic to various types of cells is one good measure for estimating how much of the dispersant it would take to cause cell death. The more dispersant it takes to cause cell death, the less toxic the dispersant. Estrogen and androgen receptors are proteins in the body that interact with the hormones estrogen and testosterone and respectively control development and function of the female (estrogen) and male (androgen) reproductive organs.

**Q9. What tests did EPA use to assess the degree each dispersant is toxic to living cells (cytotoxicity)?**

**A:** *In vitro* assays are fast, often automated chemical screening tests that assess the potential for a chemical to affect specific biological processes that could impact human health and the environment.

**Q10. What are the dispersant test results for potential endocrine activity, cytotoxicity and acute toxicity to shrimp and small fish?**

**A:** While the dispersant products alone – not mixed with oil – have roughly the same effects, JD-2000 and Corexit 9500 proved to be the least toxic to small fish, and JD-2000 and SAFRON GOLD were the least toxic to the mysid shrimp.

None of the eight dispersants tested displayed biologically significant endocrine disrupting activity, with the exception of a weak response for two of the dispersants (Nokomis 3 -F4 and ZI- 400) in one of the tests. This estrogenic result is likely not of biological significance. Cell death (degree the dispersant is toxic to living cells) was observed in some tests at concentrations above 10 parts per million. The endocrine and the cytotoxicity screening were conducted at dispersant concentrations from 0.00 1 parts per million up to 10,000 parts per million.

None of the dispersants triggered cell death at the likely concentrations of dispersants expected in the Gulf.

## Toxicity Testing - Question & Answers

**Q11. Why are toxicity tests used?**

**A:** Toxicity tests are used to determine the potential adverse effects of chemicals on humans and other organisms. Acute toxicity tests are used on different species to determine lethal concentrations. Chronic toxicity tests can be used to determine sublethal chemical concentrations such as those adversely affecting reproduction, growth and developmental processes as a result of long-term (chronic) exposure. *In vitro* assays are used to screen a large quantity of chemicals using many tests to prioritize which chemicals have the potential to be the most toxic.

**Q12. What are "*in Vitro* Assay" screenings and why are they used?**

**A:** *In vitro* assays are fast, often automated chemical screening tests that do not use live animals to assess the potential of a chemical to affect specific biological processes that may impact human health and the environment. On average, it would take a researcher eight hours a day, five days a week, for 12 years to do these assays. With computer we can do these tests and get results in three days. These types of tests are used regularly in the pharmaceutical industry to study the effects of drugs and medications and are also used to assess environmental chemicals. Traditional chemical toxicity testing (typically animal tests in the lab) are time consuming and expensive. *In vitro* assay screening provides fast, often automated screening results for assessing the potential of a chemical to affect human health and the environment. This type of screening can be used to prioritize which chemicals need further toxicity testing using animals. Animal tests on one chemical can costs millions of dollars. In comparison, numerous *in vitro* tests can be run on a chemical for about $20,000.

In: Dispersants in Oil Spills: Issues and Research ISBN: 978-1-61209-241-6
Editors: Edward K. Holm 

*Chapter 6*

# DISPERSION OF CRUDE OIL AND PETROLEUM PRODUCTS IN FRESHWATER

***Brian A. Wrenn***

## FOREWORD

The U.S. Environmental Protection Agency (EPA) is charged by Congress with protecting the Nation's land, air, and water resources. Under a mandate of national environmental laws, the Agency strives to formulate and implement actions leading to a compatible balance between human activities and the ability of natural systems to support and nurture life. To meet this mandate, EPA's research program is providing data and technical support for solving environmental problems today and building a science knowledge base necessary to manage our ecological resources wisely, understand how pollutants affect our health, and prevent or reduce environmental risks in the future.

The National Risk Management Research Laboratory (NRMRL) is the Agency's center for investigation of technological and management approaches for preventing and reducing risks from pollution that threaten human health and the environment. The focus of the Laboratory's research program is on methods and their cost-effectiveness for prevention and control of pollution to air, land, water, and subsurface resources; protection of water quality in public water systems; remediation of contaminated sites, sediments and ground water; prevention and control of indoor air pollution; and restoration of ecosystems. NRMRL collaborates with both public and private sector partners to foster technologies that reduce the cost of compliance and to anticipate emerging problems. NRMRL's research provides solutions to environmental problems by: developing and promoting technologies that protect and improve the environment; advancing scientific and engineering information to support regulatory and policy decisions; and providing the technical support and information transfer to ensure implementation of environmental regulations and strategies at the national, state, and community levels.

This publication has been produced as part of the Laboratory's strategic long-term research plan. It is published and made available by EPA's Office of Research and Development to assist the user community and to link researchers with their clients.

Sally Gutierrez,

Director National Risk Management Research Laboratory

## Notice

The U.S. Environmental Protection Agency through its Office of Research and Development funded the research described here under order number EP06C000259 to Brian A. Wrenn at the Washington University Department of Energy, Environmental, and Chemical Engineering. It has been subjected to the Agency`s review and has been approved for publication as an EPA document.

## 1. Abstract

The objective of this research was to investigate the relationship between dispersion effectiveness in freshwater and the surfactant composition for fresh and weathered crude oil. Although limited research on the chemical dispersion of crude oil and petroleum products in freshwater has been conducted, previous studies did not identify the dispersants that were investigated, much less describe the chemistry of the surfactants that were used. The absence of information on surfactant composition is a major impediment to the scientific investigation of dispersant effectiveness because this information is necessary for the development of a more fundamental understanding of dispersant effectiveness. Therefore, the relationship between surfactant chemistry and dispersant effectiveness was systematically evaluated. This report showed that, at least with Mars Blend crude oil in simulated lake water, dispersants can be designed to drive an oil slick into the freshwater column with the same efficiency as in saltwater as long as the hydrophilic-lipophilic balance (HLB) is optimum. Clearly, many more oils would need to be tested under different conditions (temperature, organic content, water composition, etc.) to enable firm conclusions that oil can be dispersed in freshwater as a response tool. The ultimate decision to use dispersants in treating freshwater petroleum oil spills is up to the federal on-scene coordinator, the incident command team, the regional response teams, and EPA Headquarters, since many other factors need to be considered before rendering a decision to disperse oil into the water column. It is beyond the scope of this report to consider such factors. Its purpose was simply to determine if freshwater dispersion is possible and to determine whether effective freshwater dispersants can be designed to produce stable oil droplet distributions in such an environment.

# 2. Introduction

## 2.1. Question Addressed by This Report

This report attempts to answer the question about whether petroleum oils can be dispersed in a freshwater environment. Most dispersants currently on the National Contingency Plan Product Schedule (NCPPS) have been developed for saltwater environments, and the literature is lacking in regards to freshwater dispersion. Research data presented herein will show that effective freshwater dispersants can be developed as a response tool, but the decision to use them is a policy decision that is not addressed by this technical report.

## 2.2. Chemical Properties of Dispersants

One response alternative for marine crude petroleum oil spills is chemical dispersion. Dispersion is accomplished by addition of chemicals that interact with the floating oil to reduce the oil-water interfacial tension and facilitate the formation of small oil droplets (NRC, 2005). The individual chemicals are formulated into mixtures that are collectively known as dispersants. The composition of most commercially available dispersants is proprietary, but in general they consist of one or more nonionic surfactants dissolved in a solvent carrier (NRC, 1989). Some dispersants also include one or more anionic surfactants and other additives (*e.g.*, phosphoric acid). The purpose of the solvent is to provide the surfactant mixture with an appropriate viscosity, which ensures that it can be pumped through spray nozzles at environmental temperatures. The solvent may be water miscible (*e.g.*, 2-butoxyethanol) or immiscible (normal alkanes) (NRC, 2005).

Surfactants are the active (*i.e.*, interfacial tension reducing) ingredients in dispersants. Surfactants are compounds that have hydrophobic and hydrophilic components within the same molecule. The amphiphilic character of surfactants causes them to accumulate at interfaces because the hydrophilic part of the molecule interacts strongly with water, and the hydrophobic part interferes with more thermodynamically-favorable hydrogen bonding interactions between water molecules. These interactions are thus displaced from the water phase (Porter, 1991). Nonionic surfactants are most common in dispersants because they have much lower aqueous solubility than do anionic surfactants (Porter, 1991), and they are generally less toxic and less affected by electrolyte concentration than are anionic and cationic surfactants (Porter, 1991; Myers, 2006). This is not necessarily true for the alkyl phenol ethoxylates, which have been linked to endocrine disrupting activity. The latter surfactants were not a part of this study. The lower water solubility tends to increase the extent to which the surfactants partition into the oil phase when the aqueous phase is in great excess, as it is for environmental applications. The nonionic surfactants used in past dispersant studies are ethoxylated derivatives of fatty acids, fatty alcohols, and fatty acid esters of sorbitan (NRC, 1989; Fingas *et al.,* 1990; Georges-Ares and Clark, 2000), in which the hydrophilic portion of the molecule consists of polyethylene glycol chains of varying lengths, and the hydrophobic portion is contributed by the fatty acyl chains, usually ranging between about 12 and 18 carbon atoms in length.

A typical measure of the relative importance of the hydrophobic and hydrophilic portions of nonionic surfactants is the hydrophile-lipophile balance (HLB), which ranges from zero for completely lipophilic (hydrophobic) molecules to 20 for completely hydrophilic (uncharged) molecules. Packing arguments suggest that the dominant group, characterized by the ratio of linear cross-sectional areas, will tend to orient into the continuous phase (NRC, 1989; Porter, 1991). So, surfactants with low HLB tend to stabilize water-in-oil emulsions, whereas those with high HLB stabilize the more desirable oil-in-water emulsions (NRC, 1989; Clayton *et al.,* 1993).

Commercial dispersants tend to have overall HLBs in the range of 9 to 11, which is often achieved by combining surfactants with higher and lower HLB. Although the industry consensus suggests that combining surfactants with different HLB improves dispersant effectiveness (NRC, 1989; Clayton *et al.*, 1993), others have offered alternative findings (Fingas *et al.*, 1990).

Although the HLB is the parameter that is most commonly used to describe the characteristics of dispersants, the same HLB can be obtained in many different ways. For example, a single surfactant with the desired HLB can be used, or two or more surfactants can be mixed in proportions that give the same desired HLB. The HLB of surfactant mixtures is given by the mass-weighted average of the individual surfactants (Myers, 2006). Furthermore, dispersants with different chemical characteristics may have the same HLB but exert different effects on the system (Bruheim *et al.*, 1999; Van Hamme and Ward, 1999; Bruheim and Eimhjellen, 2000). Therefore, the composition of the dispersant formulations used in this research were varied systematically to produce dispersants with similarly defined HLBs but different chemical compositions.

## 2.3. Effect of Salinity on Dispersion Effectiveness

The effectiveness of chemical dispersants can be strongly influenced by salinity (*i.e.*, ionic strength), but the relationship between dispersion effectiveness and salinity can vary for different dispersant-oil combinations (Lehtinen and Vesala, 1984; Belk *et al.*, 1989; Fingas *et al.*, 1991; Blondina *et al.*, 1999). Dispersants that were optimized for marine use [*e.g.*, Corexit 9500 (http://www.epa.gov/emergencies/docs/oil/ncp/schedule.pdf), Enersperse 700 (no longer on the EPA Product Schedule)] are often ineffective in freshwater (Fingas *et al.*, 1991; Blondina *et al.*, 1999), while those that are optimized for use in freshwater are less sensitive to salinity (Belk *et al.*, 1989). Although most studies vary salinity by diluting natural or artificial seawater, which is dominated by sodium and chloride ions, minor ions such as calcium and magnesium may be more important in determining dispersion effectiveness. One study using the Labofina method of measuring dispersant effectiveness showed very high effectiveness of freshwater and marine dispersants at calcium concentrations that were low relative to seawater concentrations (Belk *et al.*, 1989). The effectiveness of the unidentified marine dispersant increased from 6% in deionized water to 81% in water containing calcium at a concentration of 400 mg/L as $CaCO_3$, which is less than half the concentration found in seawater but still high relative to most freshwater. Dispersants optimized for freshwater were less sensitive to the calcium concentration than those optimized for marine use.

# 3. Objective of Research

The objective of this research was to investigate the relationship between dispersion effectiveness in freshwater and the surfactant composition for weathered crude oil. Although limited research on the chemical dispersion of crude oil and petroleum products in freshwater has been conducted, previous studies (*e.g.*, Belk *et al.,* 1989) did not identify the dispersants that were investigated, much less describe the chemistry of the surfactants that were used. The absence of information on surfactant composition is a major impediment to the scientific investigation of dispersant effectiveness because this information is necessary for the development of a more fundamental understanding of dispersant effectiveness. Therefore, the relationship between surfactant chemistry and dispersant effectiveness was systematically evaluated.

# 4. Materials and Methods

## 4.1. Surfactants

The surfactants that were used in this research are listed in Table 1. The surfactants were selected from three chemical classes: sorbitan esters, fatty acid ethoxy esters, and fatty acyl ethoxy ethers. Because many of the surfactants used in this research are widely distributed and used for a variety of purposes, Table 1 provides the trade names in addition to the chemical names of the surfactants that were used. For brevity, the structure of the hydrophilic groups is given as POE(x), where "POE" means "polyoxyetheylene" and "x" is the total number of ethylene oxide units in the POE chains. The size of the hydrophilic groups is also abbreviated as "$E_x$" in some abbreviations. The meaning of the variable 'x' in $E_x$ is identical to POE(x). They both refer to the number of ethoxyl groups. (The sorbitan ester surfactants frequently have several POE chains esterified to different hydroxyl groups of the sorbityl group, but the total number of ethylene oxide monomers is as indicated.) Example of polyoxyethylene (20) sorbitan monolaurate is shown below:

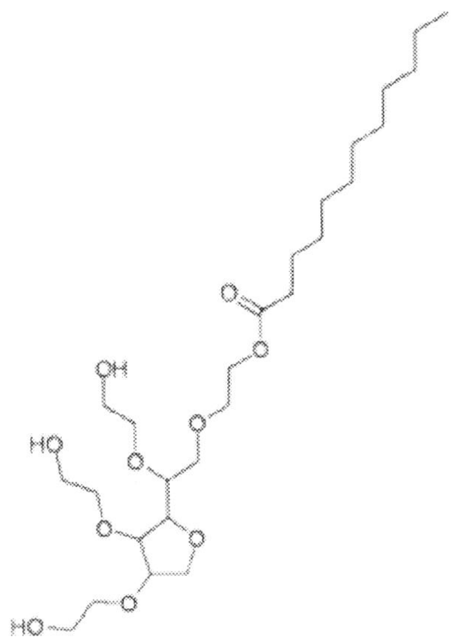


The chain length and chemical characteristics of the fatty acyl groups is indicated by $C_{y:z}$, where "y" is the number of carbon atoms in the fatty acyl chain and "z" is the number of carbon- carbon double bonds. The number of carbon-carbon double bonds is important because, for a given chain length, the melting point is a function of the degree of unsaturation. Greater unsaturation (*i.e.*, number of double bonds) is correlated with lower melting point for a given chain length and, therefore, greater tendency to exist as a liquid at ambient temperatures. Several of the surfactants include more than one fatty acyl group in the hydrophobic portion. These are indicated by the use of "di-" (*i.e.*, two fatty acyl groups), "tri-" (*i.e.*, three fatty acyl groups), or "hexa-" (*i.e.*, six fatty acyl groups). The Tergitols (fatty acyl ethers) are composed of mixtures of fatty acyl groups varying from 11 to 15 carbon atoms in chain length.

The HLB of the individual surfactant molecules is determined by the balance between the size of the hydrophilic group (*i.e.*, "x", the number of ethylene oxide units in the polyoxyethylene chain) and the length (*i.e.*, "y", the number of carbon atoms) and number (*i.e.*, di-, tri-) of the fatty acyl group that constitutes the hydrophobic portion.

**Table 1. List of Surfactant Trade Names and Chemical Names**

| Surfactant class | Trade name | Chemical name | HLB |
|---|---|---|---|
| sorbitan esters | Span 20 | sorbitan monolaurate ($S-C_{12}$) | 8.6 |
| | Span 40 | sorbitan monopalmitate ($S-C_{16}$) | 6.7 |
| | Span 60 | sorbitan monostearate ($S-C_{18}$) | 4.7 |
| | Span 80 | sorbitan monooleate ($S-C_{18:1}$) | 4.3 |
| | Tween 60 | POE(20) sorbitan monostearate ($E_{20}-S-C_{18}$) | 14.9 |
| | Tween 65 | POE(20) sorbitan tristearate ($E_{20}-S-triC_{18}$) | 10.5 |
| | Tween 80 | POE(20) sorbitan monooleate ($E_{20}-S-C_{18:1}$) | 15.0 |
| | Tween 85 | POE(20) sorbitan trioleate ($E_{20}-S-triC_{18:1}$) | 11.0 |
| | | POE(50) sorbitan hexaoleate ($E_{50}-S-hexaC_{18:1}$) | 11.6 |
| fatty acid ethyoxy esters | | POE(2) monolaurate ($E_2-C_{12}$) | 6.5 |
| | | POE(4) monolaurate ($E_4-C_{12}$) | 9.8 |
| | | POE(12) monolaurate ($E_{12}-C_{12}$) | 14.9 |
| | | POE(12) dilaurate ($E_{12}-diC_{12}$) | 12.2 |
| | | POE(4) monostearate ($E_4-C_{18}$) | 7.9 |
| | Myrj 45 | POE(8) monostearate ($E_8-C_{18}$) | 11.2 |
| | Myrj 52 | POE(40) monostearate ($E_{40}-C_{18}$) | 17.3 |
| | | POE(4) distearate ($E_4-diC_{18}$) | 5.0 |
| fatty alcohol ethyoxy ethers | | POE(6) tridecyl ether ($E_6-C_{13}$) | 11.4 |
| | Brij 52 | POE(2) hexadecyl ether ($E_2-C_{16}$) | 5.2 |
| | Brij 58 | POE (20) hexadecyl ether ($E_{20}-C_{16}$) | 15.7 |
| | Brij 76 | POE(10) stearyl ether ($E_{10}-C_{18}$) | 12.4 |
| | Brij 98 | POE(20) oelyl ether ($E_{20}-C_{18:1}$) | 15.3 |
| | Tergitol-15-S-3 | ($E_3-C_{11-15}$) | 8.0 |
| | Tergitol-15-S-5 | ($E_5-C_{11-15}$) | 10.0 |
| | Tergitol-15-S-7 | ($E_7-C_{11-15}$) | 12.1 |

**Table 2. Composition of Synthetic Lake Water**

| Component | Concentration, Lake Michigan | mg/l Synthetic Lake Water |
|---|---|---|
| calcium | 34 | 35 |
| magnesium | 10 | 11 |
| sodium | 5 | 26 |
| chloride | 8 | 62 |
| sulfate | 20 | 43 |
| alkalinity | 108 | 57 |

## 4.2. Composition of Synthetic Lake Water

Experiments were conducted in artificial freshwater that was designed to have a divalent cation composition similar to Lake Michigan water. The artificial Lake Michigan water has the following composition (mg/l): $NaHCO_3$ (96), $CaCl_2 \cdot 2H_2O$ (128), $MgSO_4 \cdot 7H_2O$ (111). It is not known to what extent this formulation differs from other freshwater sources. After dissolving the components, the artificial lake water was aerated to reduce the pH to approximately 8 and filtered through a 0.2 μm membrane filter. The synthetic lake water and Lake Michigan water are compared in Table 2. The alkalinity of the synthetic lake water is lower than Lake Michigan water, and the sodium and chloride concentrations are higher, but these factors were not expected to have a significant effect on dispersant performance. Natural organic matter might also have an influence on dispersant performance, but such a factor is difficult to simulate in a small laboratory study such as this.

## 4.3. Experimental Design

The dispersant formulations that were tested in this research are shown in Table 3. The dispersants were prepared by dissolving the surfactants in n-dodecane (Sigma-Aldrich, St. Louis, MO). The total concentration of surfactants in all dispersant formulations was 70% by mass, and the concentration of dodecane was 30%. These dispersant formulations were designed to test the effects of three factors on performance: (A) HLB, (B) surfactant chemical characteristics, and (C) dispersant composition. All treatment factors were tested at three levels. The dispersant formulations shown in Table 3 encompass the HLB range of most interest (*i.e.*, between about 8 and 12), and the dispersants were formulated using surfactants with three different chemical characteristics: sorbitan esters, fatty acid ethoxylates, and ethoxylated fatty acyl ethers. Some investigators have suggested that dispersants with lower HLB should be more effective in freshwater, whereas dispersants with higher HLB should be more effective in seawater (Clark, personal communication). The surfactants used to formulate these dispersants are similar in that the hydrophilic part of the molecule is composed primarily of polyethoxy groups, and the hydrophobic part is contributed by long-chain ($C_{12}$ to $C_{18}$) fatty acyl chains, which may be fatty acids (as in sorbitan ester and fatty ester surfactants) or fatty alcohols (as in fatty ether surfactants). The biggest structural difference between these surfactants is the presence of a sorbitan molecule (*i.e.*, a cyclic anhydride of the sugar alcohol sorbitol), to which the fatty acid and polyethoxy groups are connected, in the sorbitan esters. Also, the Span-type sorbitan esters are not ethoxylated. The third factor that was tested, dispersant composition, evaluated the benefit conferred by using mixtures of two surfactants instead of a single surfactant. Since the superiority of surfactant mixtures over single-surfactant formulations has been rationalized based on packing arguments (Porter, 1991; Myers, 2006), it seems reasonable to speculate that mixtures that are dominated by one surfactant will not be as effective as those that contain equal amounts of all surfactants. Therefore, dispersants with the same HLB were prepared in three different ways for each surfactant chemical class: one dispersant was prepared with a single surfactant with HLB that was sufficiently close to the target HLB, one dispersant was prepared by adding two surfactants in different proportions, and the third was prepared by combining two surfactants in similar proportions (*i.e.*, close to a 1:1 ratio).

**Table 3. Dispersant Formulations for Testing in Freshwater**

| HLBtarget | | Surfactant Type[†] | | |
|---|---|---|---|---|
| | | **Sorbitan Esters (B1)** | **Fatty Acid Ethoxy Esters (B2)** | **Fatty Alcohol Ethoxy Ethers (B3)** |
| 8 (A1) | C1 | Span 20 | $E_4$-$C_{18}$ | Tergitol 15-S-3 |
| | C2 | 65% Span 80 + 35% Tween 80 | 70% di$C_{18}$-$E_4$ + 30% $E_{12}$-$C_{12}$ | 73% Brij 52 + 27% Brij 98 |
| | C3 | 48% Span 60 + 52% Tween 85 | 54% $E_4$-di$C_{18}$ + 46% Myrj45 | 56% Brij 52 + 44% E6 -C13 |
| 10 (A2) | C1 | Tween 65 | $E_4$-$C_{12}$ | Tergitol 15-S-5 |
| | C2 | 60% Span 40 + 40% Tween 60 | 31% $E_2$-$C_{12}$ + 69% Myrj45 | 34% Brij 52 + 66% Brij 76 |
| | C3 | 47% Span 80 + 53% Tween 80 | 50% $E_4$-di$C_{18}$ + 50% $E_{12}$-$C_{12}$ | 53% Brij 52 + 47% Brij 98 |
| 12 (A3) | C1 | $E_{50}$-S-hexa$C_{18:1}$ | $E_{12}$-di$C_{12}$ | Tergitol 15-S-7 |
| | C2 | 28% Span 80 + 72% Tween 80 | 29% $E_4$-di$C_{18}$ + 71% $E_{12}$-$C_{12}$ | 33% Brij 52 + 67% Brij 98 |
| | C3 | 46% Span 20 + 54% Tween 60 | 49% $E_2$-$C_{12}$ + 51% Myrj52 | 48% Tergitol 15-S-3 + 52% Brij 58 |

see Table 1 and Section 3.1 for description of the chemical characteristics of these surfactants and for a definition of the abbreviations

The only commercial dispersant whose composition is known is Corexit 9500 with an HLB of 10-11, and its composition is similar to the Span80-Tween80 ratio shown in Table 3 under Sorbitan esters for HLB 10.

A total of 10 g of each dispersant (7 g of surfactants plus 3 g n-dodecane) was prepared. Each component was weighed to an accuracy of + 0.01 g. After the surfactants and solvent were combined, the mixtures were heated (if required to make the surfactants dissolve) and mixed for a minimum of 15 minutes at 200 RPM on an orbital shaker. Although all formulations were evaluated, not all have physical characteristics that would be conducive to practical application in oil-spill response. Two problems were noted with these formulations: (1) some of the dispersants were solids at room temperature, and (2) some of the surfactants were insoluble in the solvent used to make the dispersant (dodecane). The physical characteristics of the dispersants are indicated in Table 4, and the problem formulations are depicted in boldface font. the problem formulations are depicted in boldface font

Dispersant formulations that were solids at room temperature were equilibrated at 50 $^{o}$C prior to use to facilitate their use in dispersion experiments. All of the dispersant formulations that were tested in this research were liquids at this temperature. To prevent refreezing of the dispersants during transfer, the pipette tips were also equilibrated at 50 $^{o}$C before use. Dispersants for which one or more of the surfactants was insoluble in the solvent were well mixed prior to use. This procedure appeared to be adequate for use in these dispersion experiments because it produced a cloudy suspension that separated into two phases relatively slowly.

**Table 4. Physical Characteristics of Dispersant Formulations**

| **HLBtarget** | | **Sorbitan Esters (B1)** | **Fatty Acid Ethoxy Esters (B2)** | **Fatty Alcohol Ethoxy Ethers (B3)** |
|---|---|---|---|---|
| 8 (A1) | C1 | homogenous liquid | **2-phase solid** | homogenous liquid |
| | C2 | homogenous liquid | homogenous **solid** | homogenous **solid** |
| | C3 | homogenous **solid** | homogenous **solid** | 2-phase liquid |
| 10 (A2) | C1 | **2-phase solid** | homogenous liquid | homogenous liquid |
| | C2 | homogenous solid | **2-phase solid** | homogenous **solid** |
| | C3 | homogenous liquid | homogenous **solid** | homogenous **solid** |
| 12 (A3) | C1 | homogenous liquid | homogenous liquid | homogenous liquid |
| | C2 | homogenous liquid | **2-phase solid** | homogenous **solid** |
| | C3 | homogenous liquid | homogenous solid | homogenous solid |

Each dispersant formulation was tested using three independently replicated effectiveness tests. The order of testing was randomized except that all dispersants were tested once before any dispersant was tested a second time. Because it took several weeks to test all of the dispersants once, this type of blocking prevented confounding of treatment effects with time effects. That is, in a completely randomized design, it would have been possible for all three replicates of one dispersant to be conducted at almost the same time and all three replicates of another dispersant to be conducted at another time. If slow systematic changes in performance occurred over time, the effect of time could have been identified as a treatment effect. Such temporal grouping of replicates was not possible with the experimental design that was used.

## 4.4. Oil

Artificially weathered Mars Blend crude oil (from the Gulf of Mexico) was used in these experiments. Approximately 200 ml of this oil, which had an initial API gravity of about 30, was weathered by evaporation under nitrogen for 3 days. Density measurements were not made after weathering. To minimize mass transfer limitations, the oil was mixed by magnetic stirring during the evaporation process. The mass of oil was reduced by 18.5% by this process. This oil is listed in Environment Canada's ETC database.

## 4.5. Dispersion Effectiveness Experiments

The dispersant formulations listed above in Table 3 were tested using the baffled flask test (BFT), a bench-scale dispersion effectiveness test that was developed for testing of dispersants for listing on the National Contingency Plan Product Schedule (Sorial *et al.*, 2004). Dispersion effectiveness was evaluated by adding 100 μl of weathered Mars Blend crude oil to a 150-ml baffled flask containing 120 ml of synthetic freshwater using a Repeater Plus Pipette (Eppendorf, Westbury, NY) with a 0.5 ml pipette tip. Next, 4 μl of dispersant was added to the floating oil using the same pipette with a 0.1 ml pipette tip. This produced a dispersant-to-oil ratio (DOR) of 1:25. This baffled flask was mixed for 10 minutes at 200 rpm using a Lab-Line Orbit Environ Shaker (Lab-Line Instruments, Inc., Melrose Park, IL). After the mixing period, the flask was carefully removed from the shaker and the dispersion was allowed to settle for 10 minutes. Figure 1 shows the baffled flask during the settling period.

After the settling period, a 40-ml sample was collected through the stopcock at the bottom of the baffled flask. A 30-ml subsample of the 40-ml sample was extracted with dichloromethane (DCM) to measure the mass of dispersed oil, and a 40-μl sample from the remaining volume was diluted with 100 ml of deionized water and analyzed using a model LS- 200 Liquilaz Optical Particle Counter (OPC) (Particle Measuring Systems, Inc, Boulder, CO) to determine the size distribution of dispersed oil droplets.

Dispersion effectiveness was estimated from the mass concentration of dispersed oil, which was measured by extraction of the oil into DCM. A 30-ml sample of the dispersed phase was added to a 125-ml separatory funnel containing 5 ml of a 250 g/L sodium chloride solution. The dispersed oil was then extracted by adding 10 ml of DCM to the separatory funnel, and the mixture was shaken for approximately two minutes. This mixture was allowed to settle until it separated into two distinct phases, and the DCM phase was collected by draining from the bottom of the separatory funnel. This procedure was repeated at least three times and until the DCM phase was colorless after extraction. Addition of sodium chloride facilitated separation of the DCM and aqueous phases into two separate phases with a distinct interface.

Figure 1. Baffled flask during the settling period.

The solvent recovered from all three extraction steps was combined and filtered through DCM-rinsed anhydrous sodium sulfate to remove water. The resulting solution was transferred to a tared weighing-vial and the solvent was evaporated under a stream of air in a N-Evap Analytical Evaporator (Organomation Associates, Berlin, MA). After evaporation of the DCM, the tared vials were placed in a desiccator for 24 hours to remove any residual moisture. The vials were then weighed a second time to determine the mass of the dispersed oil.

The dispersion effectiveness (,) was estimated from the mass of the extracted oil as follows:

$$\eta = \frac{(M_f - M_o)\left(\frac{V_{aq\ tot}}{V_{aq\,extract}}\right)}{V_{oil,tot}\,\rho_{oil}} \times 100\% \qquad (1)$$

where $M_o$ is the initial mass of the tared vial, Mf is the mass of the vial containing the residue remaining after evaporating the solvent, $V_{aq,tot}$ is the total aqueous-phase volume used in the dispersion experiments (*i.e.*, 120 ml), $V_{aq,extract}$ is the volume of aqueous phase that was extracted (*i.e.*, 30 ml), $V_{oil,tot}$ is the volume of oil added at the beginning of the experiment (*i.e.*, 0.1 ml), and $\rho_{oil}$ is the density of the weathered Mars Blend crude oil (*i.e.*, 0.9 142 g/ml). The oil density was measured by weighing at least 10 aliquots of oil delivered by the Eppendorf Repeater Plus Pipette fitted with the 0.5-ml tip.

The dispersion efficiency and the size distribution of the dispersed oil droplets was also measured using an optical particle counter (OPC). The OPC counts the number of oil droplets within 15 user-defined size ranges between 2 and 120 ìm. The droplet-size distribution was used to estimate the total volume of dispersed oil, which was used to calculate the dispersion efficiency, and two particle-size statistics: the number mean diameter (NMD, $\bar{d}$ ) and the diameter of mean volume (DMV, $d_{\bar{V}}$ ). Six replicate measurements of the number of oil droplets in each size range ($n_i$) were averaged before calculating size-distribution statistics. (The droplet-size distribution was actually measured nine times for each sample, but the first three measurements were discarded to insure that the detector was completely flushed of previous samples or wash solutions.) The average diameter for each size range was used to represent all droplets within the bin. For example, if the lower limit of the size bin was 2 ìm and the upper limit was 3 μm, all droplets within the bin were assumed to have a diameter of 2.5 μm.

The NMD ( $\bar{d}$, μm) is a simple average of the droplet sizes:

$$\bar{d} = \frac{\sum_{i=1}^{15} \bar{n}_i d_i}{N_{tot}} \quad (2)$$

where $d_i$ is the average diameter of size bin "i", $\bar{n}_i$ is the average number of droplets in bin "i" from six replicate measurements as described above, and $N_{tot}$ is the total number concentration of oil droplets in the sample:

$$N_{tot} = \sum_{i=1}^{15} \bar{n}_i \quad (3)$$

The DMV ( $d_{\bar{V}}$ ; μm) is the volume-weighted average droplet diameter:

$$d_{\bar{V}} = \sqrt[3]{\frac{\sum_{i=1}^{15} \bar{n}_i d_i^3}{N_{tot}}} \quad (4)$$

The DMV is a better measure of the mean of the droplet-size distribution because droplet volume is proportional to the mass, whereas number is not (*i.e.*, a small droplet contains less oil than a large droplet, but both are weighted equally in calculation of the NMD). The NMD is always smaller than the DMV in these size distributions because a large number of small droplets are formed during dispersion, but they represent a relatively small fraction of the total oil mass.

The dispersion efficiency was estimated as follows:

$$\eta = \frac{\left(\frac{\pi d_{\bar{V}}^3}{6} N_{tot}\right)\left(\frac{V_{dil}}{V_{sample}}\right) V_{aq,tot}}{V_{oil,tot}\left(10^{12}\ \mu m^3/ml\right)} \times 100\% \quad (5)$$

where $V_{dil}$ is the volume of the diluted sample (typically 100 ml), $V_{sample}$ is the volume of the dispersed oil sample collected from the baffled flask after 10 minute settling period (typically 0.04 ml), and $V_{aq,tot}$ is the total volume of the aqueous phase used in the dispersion experiments (120 ml).

The results were analyzed using analysis of variance (ANOVA) to evaluate the statistical significance of any main effects or interactions. Where significant differences were identified ($p < 0.05$) by ANOVA, the means were separated using Tukey's method for paired comparisons. ANOVA and Tukey's method were performed using Systat version 10 (SPSS, Inc., Chicago, IL).

# 5. RESULTS

## 5.1. Comparison of Dispersion Effectiveness Measurements

Two methods were used to measure dispersion effectiveness in this research: liquid- liquid extraction of the dispersed oil from the aqueous phase into dichloromethane (DCM) followed by gravimetric analysis of the extracted oil (Eq. 1) and estimation of the total volume of dispersed oil from the droplet-size distribution measured using the optical particle counter (OPC). On average, the OPC gave higher estimates of dispersion effectiveness than did the gravimetric method ($\eta_{OPC} = 76.7 \pm 10.0\%$ vs. $\eta_{grav} = 68.1 \pm 10.2\%$), but the variance of the gravimetric method was smaller (avg. $s_{OPC} = 8.8 \pm 3.6\%$ vs. avg. $s_{grav} = 5.0 \pm 3.4\%$). The average dispersion effectiveness (averaged over the three independent replicate dispersion experiments) as measured by both methods is shown in Table 5. The average difference between the OPC-based effectiveness and the effectiveness estimated by the gravimetric method was $8.6 \pm 12.7\%$ [degrees of freedom (d.f.) = 26], which was statistically significant ($p = 0.002$). The lower average dispersion effectiveness measured by the gravimetric method was probably due to evaporation of volatile oil components during evaporation of the solvent. Loss of extracted oil by evaporation was observed to occur despite using evaporatively weathered crude oil in these experiments. Due to the systematic underestimation of dispersion effectiveness by the gravimetric method, the treatment effects associated with dispersant formulation were evaluated using the OPC-based measurement of dispersion effectiveness.

**Table 5. Comparison of Dispersion Effectiveness Using OPC and Gravimetric Analysis**

| HLB | Chemistry | Composition | dispersion effectiveness OPC | (%) gravimetric |
|---|---|---|---|---|
| 8 | sorbitan ester | single | 72.7 ± 6.18 | 79.7 ± 3.40 |
| | | two, unequal | 86.2 ± 3.03 | 73.3 ± 0.67 |
| | | two, equal | 96.2 ± 7.07 | 68.1 ± 7.74 |
| | fatty acid ester | single | 82.2 ± 10.6 | 68.2 ± 7.74 |
| | | two, unequal | 80.3 ± 4.04 | 63.8 ± 2.71 |
| | | two, equal | 78.6 ± 16.8 | 60.6 ± 6.53 |
| | fatty alcohol ether | single | 74.4 ± 6.60 | 83.4 ± 2.16 |
| | | two, unequal | 79.4 ± 8.01 | 68.5 ± 2.64 |
| | | two, equal | 71.6 ± 9.28 | 83.2 ± 6.49 |
| 10 | sorbitan ester | single | 82.8 ± 12.0 | 60.2 ± 16.9 |
| | | two, unequal | 93.7 ± 6.32 | 79.9 ± 1.38 |
| | | two, equal | 78.4 ± 13.0 | 68.4 ± 4.38 |
| | fatty acid ester | single | 79.0 ± 4.03 | 83.8 ± 3.54 |
| | | two, unequal | 72.4 ± 8.48 | 59.6 ± 8.31 |
| | | two, equal | 79.3 ± 11.9 | 58.6 ± 8.31 |
| | fatty alcohol ether | single | 71.6 ± 9.28 | 82.5 ± 4.59 |
| | | two, unequal | 65.0 ± 13.1 | 66.8 ± 1.66 |
| | | two, equal | 69.6 ± 7.09 | 63.6 ± 7.14 |
| 12 | sorbitan ester | single | 96.2 ± 5.61 | 65.6 ± 6.47 |
| | | two, unequal | 61.1 ± 15.4 | 53.6 ± 5.40 |
| | | two, equal | 79.3 ± 7.91 | 75.4 ± 0.25 |
| | fatty acid ester | single | 74.7 ± 13.9 | 61.2 ± 5.30 |
| | | two, unequal | 75.4 ± 6.03 | 61.1 ± 3.51 |
| | | two, equal | 57.4 ± 11.4 | 44.5 ± 4.40 |
| | fatty alcohol ether | single | 58.0 ± 8.04 | 76.8 ± 3.28 |
| | | two, unequal | 66.5 ± 4.73 | 56.0 ± 7.23 |
| | | two, equal | 82.4 ± 6.28 | 72.9 ± 2.49 |

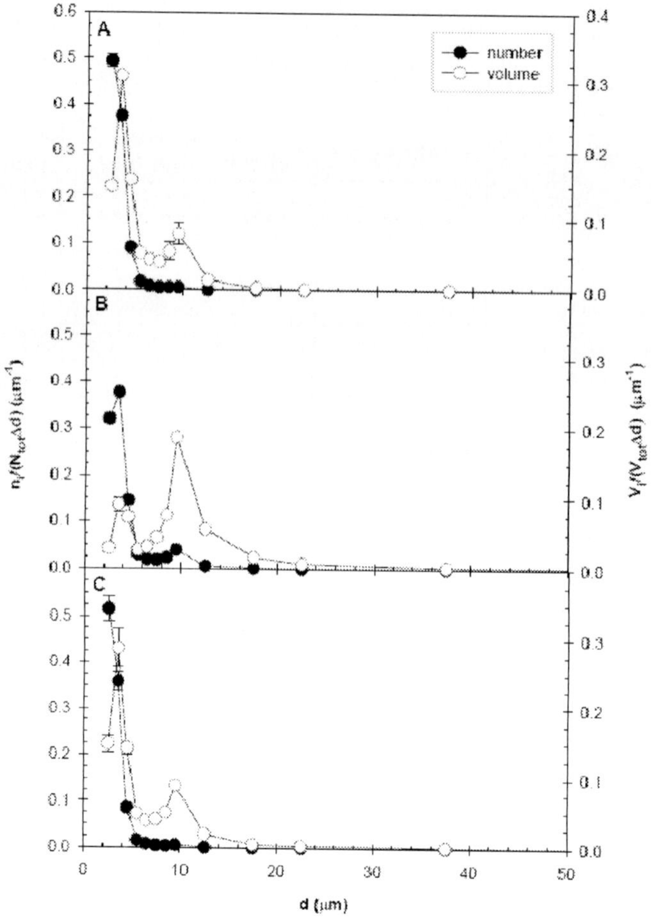


Figure 2. Comparison of normalized number (left axis) and volume (right axis) distributions for dispersions produced by single-surfactant dispersants with HLB = 8. The dispersants used included (A) Span 20 (HLB = 8.6), (B) POE(4) monostearate.

## 5.2. Size Distributions

The droplet-size distributions can be based on either number or volume. Figure 2 shows a comparison of the normalized number and volume distributions for three dispersions produced by single-surfactant dispersants with HLB = 8. Normalized size distributions are obtained by dividing the number ($n_i$) or volume ($V_i$) concentration of oil droplets in each size bin by the total number ($N_{tot}$) or volume ($V_{tot}$) concentration of the dispersion and by the size of the bin ($\Delta d_i$). These two operations ensure that the area under the size-distribution curve is equal to 1.0 (*i.e.*, the normalized size distributions have properties similar to histograms). In two of three cases, the number distribution was unimodal, but all of the volume distributions were bimodal. The (HLB = 7.9), or (C) Tergitol 15-S-3 (HLB = 8.0) as the sole surfactant. mean diameter of the largest size mode was about 10 μm, whereas the mean diameter of the smallest size mode was only about 4 ìm. The bimodality of the volume distribution reflects the disproportionate influence of large droplets. Whereas oil droplets in the larger size mode constituted less than 2% of the total number of dispersed oil droplets in dispersants prepared with Span 20 and Tergitol 15-S-3, they represented about 20% of the total dispersed oil volume. Similarly, in the dispersion produced by application of the dispersant containing POE(4) monostearate, the large droplet-size mode constituted less than 10% of the total number of dispersed droplets but it represented about 40% of the total dispersed oil volume. Because the volume of dispersed oil is more important than the number of droplets from the perspective of ecological effects and performance in spill response, the dispersant formulations tested in this research will be compared based on the characteristics of the volume distributions. It is recognized, however, that smaller droplet sizes promote emulsion stability, thereby resulting in more rapid biodegradation rates due to the resultant higher surface area of the smaller droplets.

The normalized volume distributions that were produced during this research are shown in Figures 3, 4, and 5. Figure 3 shows the volume distributions that were obtained using single- surfactant dispersants, whereas Figures 4 and 5 show the droplet-volume distributions for dispersions that were produced with dispersions prepared with two surfactants present in unequal or approximately equal proportions, respectively. All three figures also compare the size distributions obtained with dispersants prepared with sorbitan ester (panel A), fatty acyl ester (panel B), and fatty acyl ether (panel C) surfactants. They also compare the size distributions produced by dispersants with HLB of 8 (filled circles), 10 (open circles), and 12 (filled squares). The error bars represent one standard deviation for three independent replicates (*i.e.*, dispersions prepared in independent experimental units at different times). These figures clearly show that the overall characteristics of the dispersions are similar regardless of the dispersant that was used: all of the dispersions show two main size modes with mean diameters of approximately 4 ìm and 10 ìm. The distribution of oil between the two modes, however, does depend on the characteristics of the dispersant. The distribution of oil between the two modes can be quantified using the diameter of mean volume (DMV; Eq. 4).

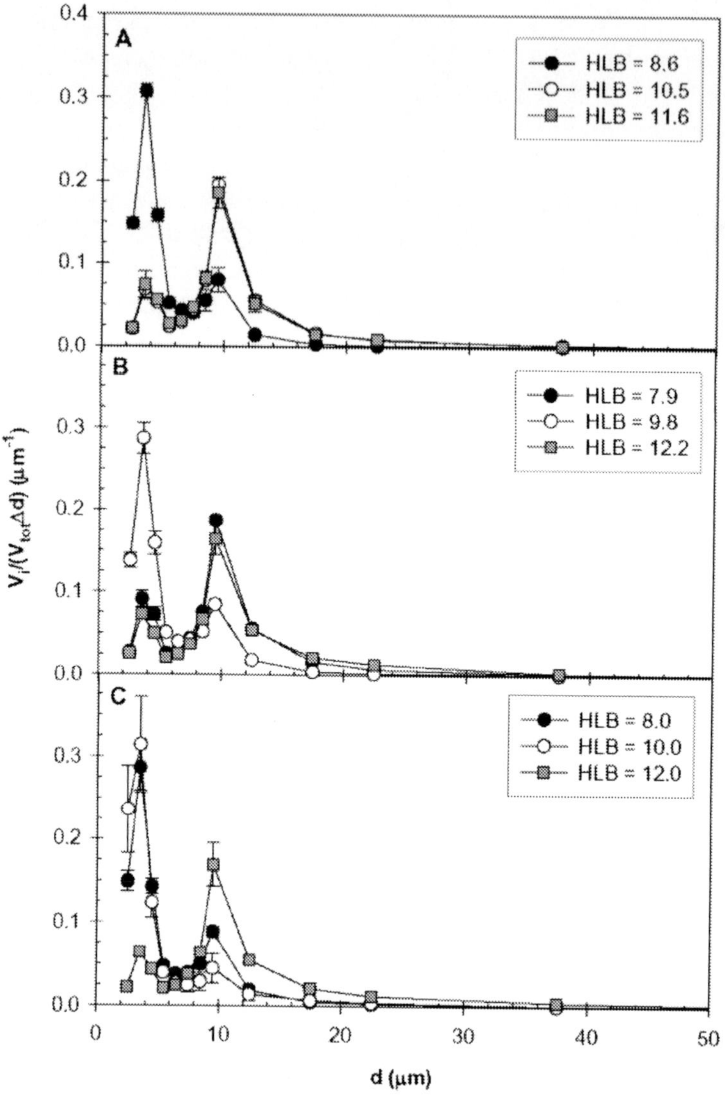


Figure 3. Normalized volume distributions for single-surfactant dispersants prepared from (A) sorbitan ester surfactants, (B) fatty acyl ester surfactants, and (C) fatty acyl ether surfactants. The error bars represent one standard deviation of three independent dispersion experiments.

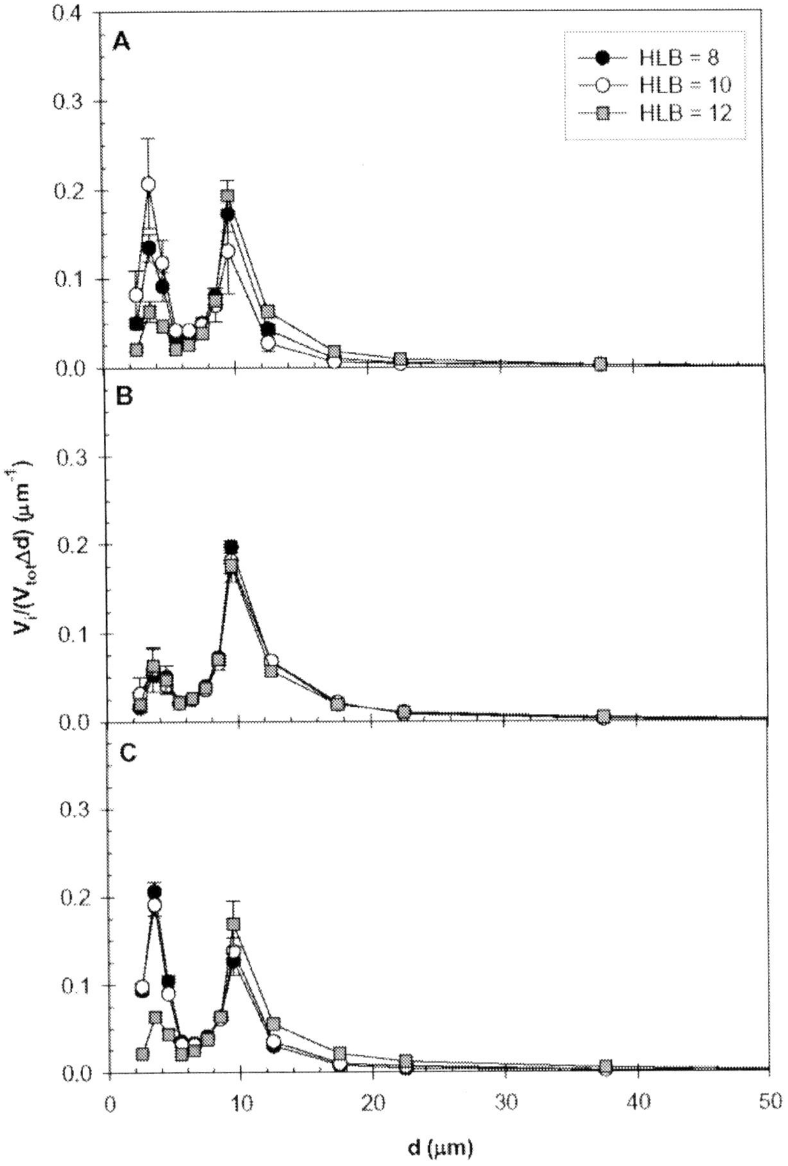


Figure 4. Normalized volume distributions for dispersions produced by dispersants prepared with two surfactants that were present in unequal proportions. The surfactants used were (A) sorbitan esters, (B) fatty acyl esters, or (C) fatty acyl ethers.

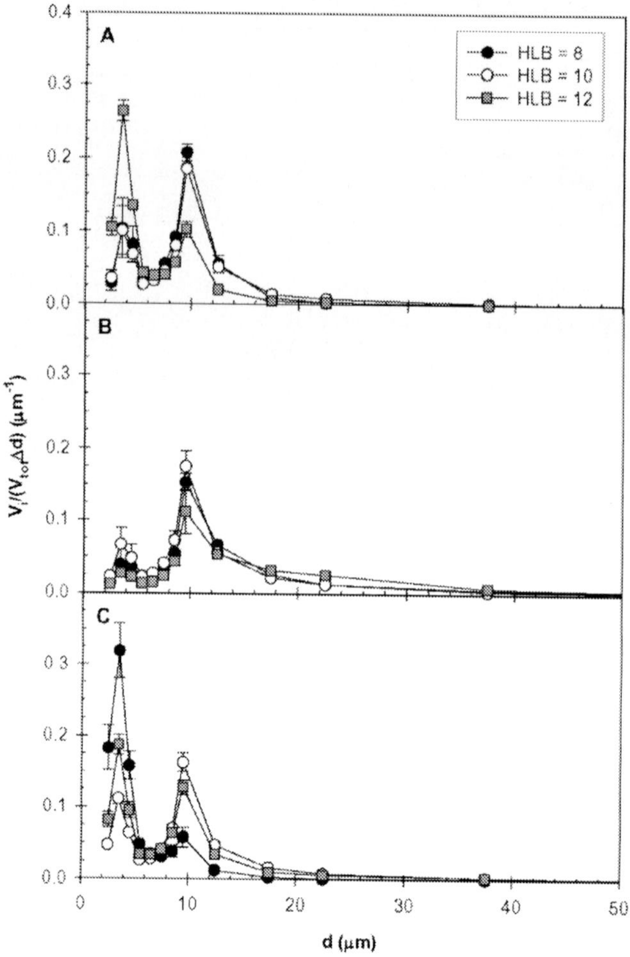


Figure 5. Normalized volume distributions for dispersions produced using dispersants prepared with two surfactants present in equal proportions. The surfactants that were used were (A) sorbitan esters, (B) fatty acyl esters, and (C) fatty acyl ethers.

## 5.3. Evaluation of Effects of Dispersant Characteristics

Treatment effects on dispersion effectiveness were evaluated by performing ANOVA on the data shown in Table 5 (OPC; based on droplet-size distribution). The null hypothesis tested was that none of the main effects or interactions was significant at the $\alpha = 0.05$ level. The results are shown in

Table 6, which shows that surfactant HLB and chemistry exerted significant effects on dispersion effectiveness. In addition, the three-factor interaction (*i.e.,* HLB, dispersant composition, surfactant chemistry) is statistically significant, indicating that the treatment main effects are not strictly additive. A similar analysis was performed to evaluate treatment effects on the droplet-size distribution, as indicated by the DMV. The results of this analysis are shown in Table 7 and demonstrate that all three main effects were highly significant ($p < 0.00\ 1$) as were all three two-factor interactions and the three-factor interaction ($p < 0.001$). means while keeping the global Type I error at 5%. In Figure 6, dispersion effectiveness means are compared only to other effectiveness means, and DMVs are compared only to other DMVs.

**Table 6. ANOVA for Treatment Effects on Dispersion Effectiveness**

| Source of Variation | Sum of Squares | Degrees of Freedom | Mean Square | F-ratio | p |
|---|---|---|---|---|---|
| HLB | 869.0 | 2 | 434.5 | 4.858 | 0.011 |
| Composition | 70.2 | 2 | 35.1 | 0.392 | 0.677 |
| Chemistry | 1743.5 | 2 | 871.8 | 9.746 | <0.001 |
| HLB x Composition | 524.5 | 4 | 131.1 | 1.466 | 0.225 |
| HLB x Chemistry | 122.6 | 4 | 30.6 | 0.343 | 0.848 |
| Composition x Chemistry | 884.9 | 4 | 221.2 | 2.473 | 0.055 |
| HLB x Composition x Chemistry | 3646.6 | 8 | 455.8 | 5.096 | <0.001 |
| Error | 4830.4 | 54 | 89.5 | | |

**Table 7. ANOVA for Treatment Effects on Diameter of Mean Volume (DMV)**

| Source of Variation | Sum of Squares | Degrees of Freedom | Mean Square | F-ratio | p |
|---|---|---|---|---|---|
| HLB | 6.68 | 2 | 3.34 | 34.18 | <0.001 |
| Composition | 11.4 | 2 | 5.71 | 58.46 | <0.001 |
| Chemistry | 44.7 | 2 | 22.4 | 228.7 | <0.001 |
| HLB x Composition | 5.58 | 4 | 1.40 | 14.28 | <0.001 |
| HLB x Chemistry | 7.02 | 4 | 1.76 | 17.95 | <0.001 |
| Composition x Chemistry | 12.0 | 4 | 3.01 | 30.78 | <0.001 |
| HLB x Composition x Chemistry | 24.3 | 8 | 3.04 | 31.08 | <0.001 |
| Error | 5.28 | 54 | 0.098 | | |

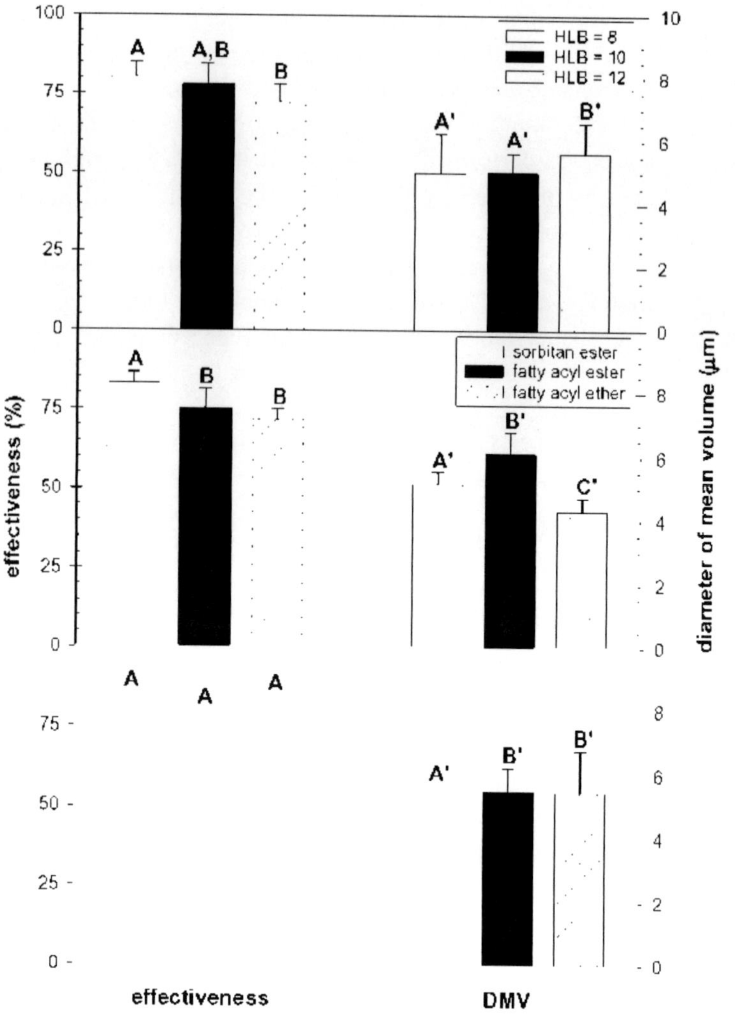


Figure 6. Comparison of treatment main effects on dispersion effectiveness (left axis) and droplet DMV (right axis). The top panel shows the effects of dispersant HLB on these performance parameters, the middle panel shows the effects of surfactant chemistry, and the bottom panel shows the effects of dispersant composition. The error bars represent one standard deviation. Bars labeled with the same letter are not significantly different at the 95% confidence level.

The treatment main effects on dispersion effectiveness and diameter of mean volume are shown in Figure 6. Significant differences between treatment means were identified using Tukey's method for pairwise comparisons, which allows comparison of all possible pairs of means while keeping the global Type I error at 5%. In Figure 6, dispersion effectiveness means are compared only to other effectiveness means, and DMVs are compared only to other DMVs.

Figure 6 shows that, whereas dispersants with HLB = 12 were less effective than those with HLB = 8, the differences between dispersants with HLB = 8 and 10 and between dispersants with HLB = 10 and 12 were not statistically significant. Dispersants with HLB = 12 also produced dispersions with a larger fraction of the dispersed oil in larger droplets. Also, dispersants that contained sorbitan-ester surfactants (*e.g.*, Tweens and Spans) were significantly more effective than those containing fatty acyl ester and fatty acyl ether surfactants. The fatty acyl ether surfactants, however, produced the smallest oil droplets. Although dispersant composition (*e.g.*, formulation with one or two surfactants) had no effect on dispersion effectiveness, dispersants formulated with a single surfactant resulted in the formation of smaller oil droplets.

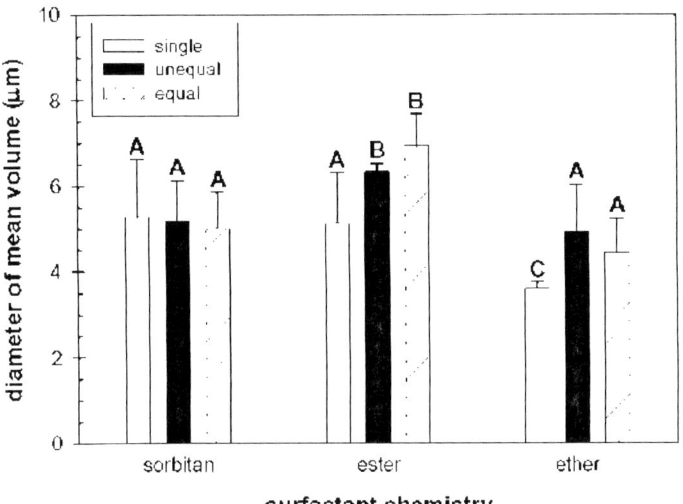


Figure 7. Comparison of treatment effects on droplet-size distribution (DMV) due to the surfactant chemistry-dispersant composition interaction. Error bars represent one standard deviation of three independent replicate dispersion experiments. Single, unequal, and equal refer to the fatty acyl chain from either a single surfactant or two surfactants in unequal or equal composition levels.

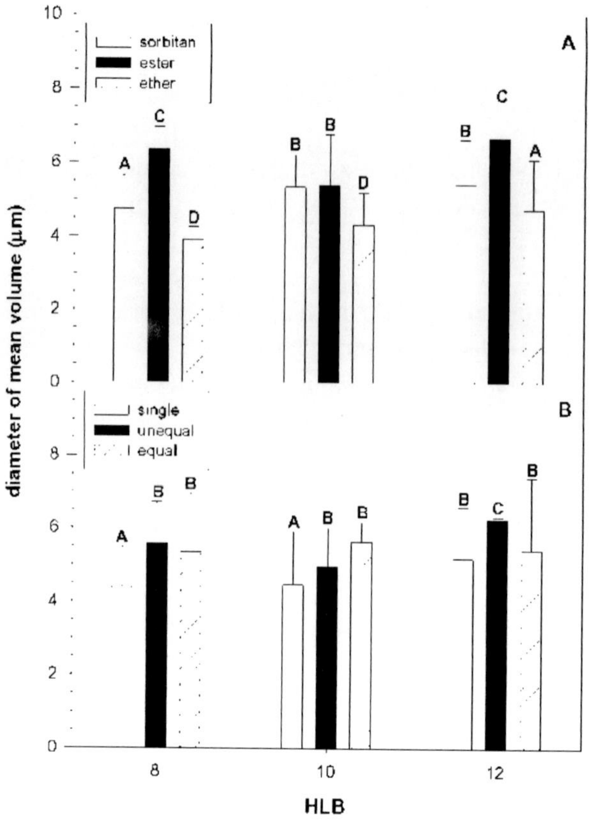


Figure 8. Comparison of treatment effects on droplet-size distribution (DMV) due to the interactions between dispersant HLB and (A) surfactant chemistry and (B) dispersant composition. Error bars represent one standard deviation of three independent replicate dispersion experiments. Single, unequal, and equal refer to the fatty acyl chain from either a single surfactant or two surfactants in unequal or equal composition levels.

Figures 7 and 8 show the effects of two-factor interactions on droplet DMV. In these figures, the significant differences that are indicated are those identified using the Least- Significant Difference (LSD) test because use of Tukey's method for pairwise comparisons produced too many comparisons (and too many differences) to easily show graphically. The differences identified using Tukey's method are shown in Tables 8-10. Figure 7 shows the interactions between surfactant chemistry and dispersant composition. In general, oil droplets formed by dispersants containing a single, fatty acyl ether surfactant were smaller than those produced with other dispersants, and droplets formed by dispersants containing two fatty acyl ester surfactants were larger. Figure 8 shows the interactions between dispersant HLB and (A) surfactant chemistry or (B) dispersant composition. In general, the smallest droplets were formed when dispersants with HLB of 8 or 10 containing fatty acyl ether surfactants were used.

**Table 8. Probabilities That the Means of Any Pair of DMVs Are Significantly Different for the Two-Factor Interaction between Dispersant HLB and Surfactant Chemistry**

| Interaction | 8-sorb | 10-sorb | 12-sorb | 8-ester | 10-ester | 12-ester | 8-ether | 10-ether | 12-ether |
|---|---|---|---|---|---|---|---|---|---|
| 8-sorb | 1.000 | | | | | | | | |
| 10-sorb | 0.005 | 1.000 | | | | | | | |
| 12-sorb | 0.001 | 1.000 | 1.000 | | | | | | |
| 8-ester | 0.000 | 0.000 | 0.000 | 1.000 | | | | | |
| 10-ester | 0.001 | 1.000 | 1.000 | 0.000 | 1.000 | | | | |
| 12-ester | 0.000 | 0.000 | 0.000 | 0.380 | 0.000 | 1.000 | | | |
| 8-ether | 0.000 | 0.000 | 0.000 | 0.000 | 0.000 | 0.000 | 1.000 | | |
| 10-ether | 0.158 | 0.000 | 0.000 | 0.000 | 0.000 | 0.000 | 0.081 | 1.000 | |
| 12-ether | 1.000 | 0.005 | 0.001 | 0.000 | 0.001 | 0.000 | 0.000 | 0.148 | 1.000 |

**Table 9. Probabilities That the Means of Any Pair of DMVs Are Significantly Different for the Two-Factor Interaction between dispersant HLB and Dispersant Composition**

| Interaction | 8-single | 8-unequal | 8-equal | 10-single | 10-unequal | 10-equal | 12-single | 12-unequal | 12-equal |
|---|---|---|---|---|---|---|---|---|---|
| 8-single | 1.000 | | | | | | | | |
| 8-unequal | 0.000 | 1.000 | | | | | | | |
| 8-equal | 0.000 | 1.000 | 1.000 | | | | | | |
| 10-single | 1.000 | 0.000 | 0.000 | 1.000 | | | | | |
| 10-unequal | 0.008 | 0.538 | 0.238 | 0.044 | 1.000 | | | | |
| 10-equal | 0.000 | 0.182 | 0.450 | 0.000 | 0.001 | 1.000 | | | |
| 12-single | 0.000 | 1.000 | 0.983 | 0.000 | 0.837 | 0.060 | 1.000 | | |
| 12-unequal | 0.000 | 0.000 | 0.000 | 0.000 | 0.000 | 0.004 | 0.000 | 1.000 | |
| 12-equal | 0.000 | 0.976 | 1.000 | 0.000 | 0.072 | 0.799 | 0.822 | 0.000 | 1.000 |

**Table 10. Probabilities That the Means of Any Pair of DMVs Are Significantly Different for the Two-Factor Interaction between Dispersant HLB and Surfactant Chemistry**

| Interaction | sorb-single | sorb-uneq | sorb-equal | ester-single | ester-uneq | ester-equal | ether-single | ether-uneq | ether-equal |
|---|---|---|---|---|---|---|---|---|---|
| sorb-single | 1.000 | | | | | | | | |
| sorb-uneq | 0.999 | 1.000 | | | | | | | |
| sorb-equal | 0.673 | 0.967 | 1.000 | | | | | | |
| ester-single | 0.988 | 1.000 | 0.995 | 1.000 | | | | | |
| ester-uneq | 0.000 | 0.000 | 0.000 | 0.000 | 1.000 | | | | |
| ester-equal | 0.000 | 0.000 | 0.000 | 0.000 | 0.005 | 1.000 | | | |
| ether-single | 0.000 | 0.000 | 0.000 | 0.000 | 0.000 | 0.000 | 1.000 | | |
| ether-uneq | 0.308 | 0.738 | 1.000 | 0.881 | 0.000 | 0.000 | 0.000 | 1.000 | |
| ether-equal | 0.000 | 0.000 | 0.005 | 0.000 | 0.000 | 0.000 | 0.000 | 0.029 | 1.000 |

The largest droplets, however, were produced by dispersants containing fatty acyl ester surfactants with HLB of 8 or 10. Overall, the correlation between dispersant effectiveness and droplet size is relatively weak, but the differences between the largest and smallest diameters of mean volume were relatively small (*i.e.*, 3.2 to 8.0 μm).

The three-factor interactions for DMV and dispersion effectiveness are shown in Figure 9A and B, respectively. Once again, due to the very large number of pairwise comparisons that are possible (351), the differences that are indicated are those identified using the LSD test. The significant differences identified using Tukey's method for pairwise comparisons are shown in Tables 11 and 12. Tukey's method identifies many more significant differences than does the LSD. For example, although the ANOVA determined that the three-factor interaction for dispersion effectiveness was highly significant ($p < 0.001$; see Table 6), the LSD test was unable to identify any differences between adjacent pairs of means when they were rank ordered, whereas Tukey's method identified 104 significant differences when all possible pairwise comparisons were evaluated (Table 12).

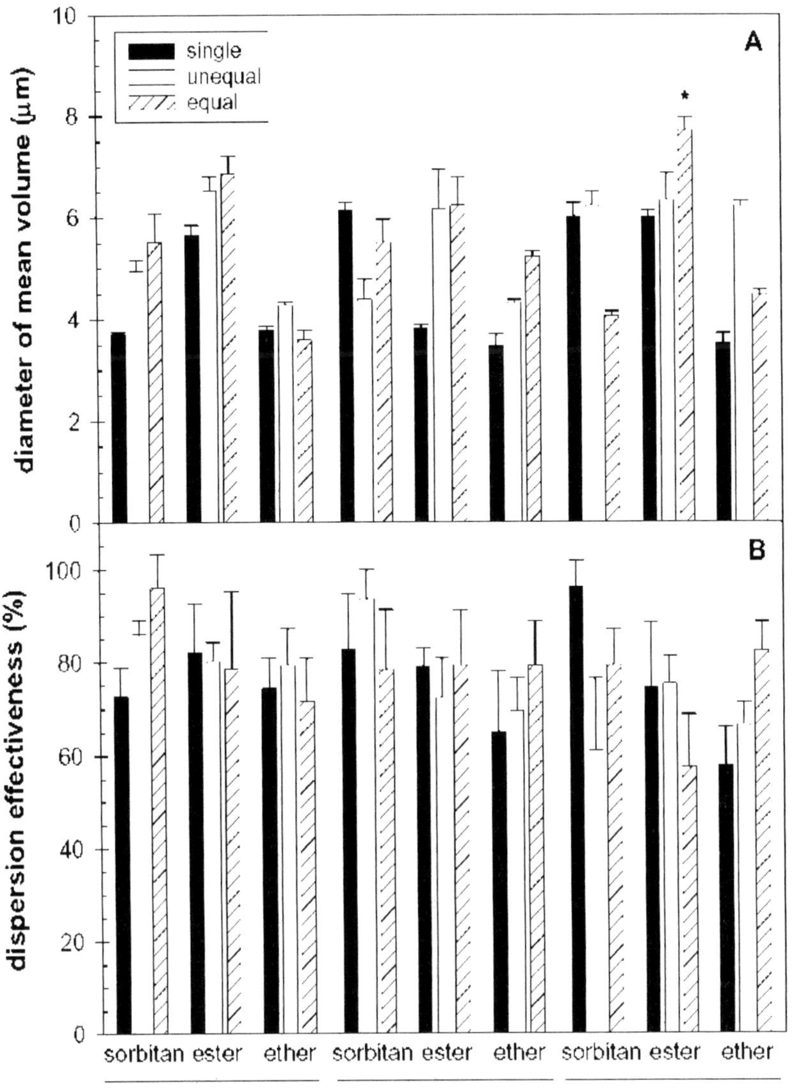


Figure 9. Comparison of three-factor interactions for (A) droplet-size distribution (DMV) and (B) dispersant effectiveness. Error bars represent one standard deviation of three independent replicate dispersion experiments. The asterisk denotes a significant difference ($p < 0.05$) as determined by the LST test.

**Table 11: Probabilities that the means of any pair of DMVs are significantly different for the three-factor interaction between dispersant HLB, surfactant chemistry, and dispersant composition**

| Interaction | 8/S/sorb | 8/U/sorb | 8/E/sorb | 10/S/sorb | 10/U/sorb | 10/E/sorb | 12/S/sorb | 12/U/sorb | 12/E/sorb | 8/S/ester | 8/U/ester | 8/E/ester | 10/S/ester | 10/U/ester | 10/E/ester | 12/S/ester | 12/U/ester | 12/E/ester | 8/S/ether | 8/U/ether | 8/E/ether | 10/S/ether | 10/U/ether | 10/E/ether | 12/S/ether | 12/U/ether | 12/E/ether |
|---|---|---|---|---|---|---|---|---|---|---|---|---|---|---|---|---|---|---|---|---|---|---|---|---|---|---|---|
| 8/S/sorb | 1.000 | | | | | | | | | | | | | | | | | | | | | | | | | | |
| 8/U/sorb | 0.003 | 1.000 | | | | | | | | | | | | | | | | | | | | | | | | | |
| 8/E/sorb | 0.000 | 0.886 | 1.000 | | | | | | | | | | | | | | | | | | | | | | | | |
| 10/S/sorb | 0.000 | 0.007 | 0.808 | 1.000 | | | | | | | | | | | | | | | | | | | | | | | |
| 10/U/sorb | 0.675 | 0.874 | 0.011 | 0.000 | 1.000 | | | | | | | | | | | | | | | | | | | | | | |
| 10/E/sorb | 0.000 | 0.921 | 1.000 | 0.753 | 0.014 | 1.000 | | | | | | | | | | | | | | | | | | | | | |
| 12/S/sorb | 0.000 | 0.027 | 0.970 | 1.000 | 0.000 | 0.951 | 1.000 | | | | | | | | | | | | | | | | | | | | |
| 12/U/sorb | 0.000 | 0.002 | 0.563 | 1.000 | 0.000 | 0.498 | 1.000 | 1.000 | | | | | | | | | | | | | | | | | | | |
| 12/E/sorb | 1.000 | 0.115 | 0.000 | 0.000 | 1.000 | 0.000 | 0.000 | 0.000 | 1.000 | | | | | | | | | | | | | | | | | | |
| 8/S/ester | 0.000 | 0.601 | 1.000 | 0.974 | 0.002 | 1.000 | 0.999 | 0.862 | 0.000 | 1.000 | | | | | | | | | | | | | | | | | |
| 8/U/ester | 0.000 | 0.000 | 0.052 | 0.998 | 0.000 | 0.040 | 0.960 | 1.000 | 0.000 | 0.165 | 1.000 | | | | | | | | | | | | | | | | |
| 8/E/ester | 0.000 | 0.000 | 0.001 | 0.443 | 0.000 | 0.001 | 0.195 | 0.701 | 0.000 | 0.003 | 1.000 | 1.000 | | | | | | | | | | | | | | | |
| 10/S/ester | 1.000 | 0.009 | 0.000 | 0.000 | 0.862 | 0.000 | 0.000 | 0.000 | 1.000 | 0.000 | 0.000 | 0.000 | 1.000 | | | | | | | | | | | | | | |
| 10/U/ester | 0.000 | 0.005 | 0.728 | 1.000 | 0.000 | 0.665 | 1.000 | 1.000 | 0.000 | 0.948 | 1.000 | 0.535 | 0.000 | 1.000 | | | | | | | | | | | | | |
| 10/E/ester | 0.000 | 0.002 | 0.535 | 1.000 | 0.000 | 0.470 | 1.000 | 1.000 | 0.000 | 0.843 | 1.000 | 0.728 | 0.000 | 1.000 | 1.000 | | | | | | | | | | | | |
| 12/S/ester | 0.000 | 0.054 | 0.973 | 1.000 | 0.000 | 0.958 | 1.000 | 1.000 | 0.000 | 0.999 | 0.997 | 0.506 | 0.000 | 1.000 | 1.000 | 1.000 | | | | | | | | | | | |
| 12/U/ester | 0.000 | 0.001 | 0.283 | 1.000 | 0.000 | 0.236 | 1.000 | 1.000 | 0.000 | 0.591 | 1.000 | 0.925 | 0.000 | 1.000 | 1.000 | 1.000 | 1.000 | | | | | | | | | | |
| 12/E/ester | 0.000 | 0.000 | 0.000 | 0.000 | 0.000 | 0.000 | 0.000 | 0.000 | 0.000 | 0.000 | 0.005 | 0.212 | 0.000 | 0.000 | 0.000 | 0.000 | 0.000 | 1.000 | | | | | | | | | |
| 8/S/ether | 0.820 | 0.516 | 0.001 | 0.000 | 1.000 | 0.001 | 0.000 | 0.000 | 1.000 | 0.000 | 0.000 | 0.000 | 0.952 | 0.000 | 0.000 | 0.000 | 0.000 | 0.000 | 1.000 | | | | | | | | |
| 8/U/ether | 0.907 | 0.619 | 0.003 | 0.000 | 1.000 | 0.004 | 0.000 | 0.000 | 1.000 | 0.001 | 0.000 | 0.000 | 0.981 | 0.000 | 0.000 | 0.000 | 0.000 | 0.000 | 1.000 | 1.000 | | | | | | | |
| 8/E/ether | 1.000 | 0.001 | 0.000 | 0.000 | 0.319 | 0.000 | 0.000 | 0.000 | 0.990 | 0.000 | 0.000 | 0.000 | 1.000 | 0.000 | 0.000 | 0.000 | 0.000 | 0.000 | 0.443 | 0.601 | 1.000 | | | | | | |
| 10/S/ether | 1.000 | 0.000 | 0.000 | 0.000 | 0.111 | 0.000 | 0.000 | 0.000 | 0.869 | 0.000 | 0.000 | 0.000 | 1.000 | 0.000 | 0.000 | 0.000 | 0.000 | 0.000 | 0.161 | 0.276 | 1.000 | 1.000 | | | | | |
| 10/U/ether | 0.850 | 0.710 | 0.004 | 0.000 | 1.000 | 0.006 | 0.000 | 0.000 | 1.000 | 0.001 | 0.000 | 0.000 | 0.960 | 0.000 | 0.000 | 0.000 | 0.000 | 0.000 | 1.000 | 1.000 | 0.507 | 0.212 | 1.000 | | | | |
| 10/E/ether | 0.000 | 1.000 | 1.000 | 0.115 | 0.249 | 1.000 | 0.297 | 0.045 | 0.007 | 0.994 | 0.001 | 0.000 | 0.000 | 0.083 | 0.040 | 0.387 | 0.013 | 0.000 | 0.057 | 0.098 | 0.000 | 0.000 | 0.134 | 1.000 | | | |
| 12/S/ether | 1.000 | 0.000 | 0.000 | 0.000 | 0.201 | 0.000 | 0.000 | 0.000 | 0.957 | 0.000 | 0.000 | 0.000 | 1.000 | 0.000 | 0.000 | 0.000 | 0.000 | 0.000 | 0.287 | 0.435 | 1.000 | 1.000 | 0.350 | 0.000 | 1.000 | | |
| 12/U/ether | 0.000 | 0.002 | 0.582 | 1.000 | 0.000 | 0.516 | 1.000 | 1.000 | 0.000 | 0.874 | 1.000 | 0.684 | 0.000 | 1.000 | 1.000 | 1.000 | 1.000 | 0.000 | 0.000 | 0.000 | 0.000 | 0.000 | 0.000 | 0.048 | 0.000 | 1.000 | |
| 12/E/ether | 0.435 | 0.974 | 0.029 | 0.000 | 1.000 | 0.037 | 0.000 | 0.000 | 0.994 | 0.007 | 0.000 | 0.000 | 0.656 | 0.000 | 0.000 | 0.000 | 0.000 | 0.000 | 1.000 | 1.000 | 0.160 | 0.046 | 1.000 | 0.452 | 0.092 | 0.000 | 1.000 |

Note: All p-values shown as 0.000 signify < 0.001.

**Table 12: Probabilities that the means of any pair of dispersant effectivenesses are significantly different for the three-factor interaction between dispersant HLB, surfactant chemistry, and dispersant composition**

| Interaction | 8/S/ sorb | 8/U/ sorb | 8/E/ sorb | 10/S/ sorb | 10/U/ sorb | 10/E/ sorb | 12/S/ sorb | 12/U/ sorb | 12/E/ sorb | 8/S/ ester | 8/U/ ester | 8/E/ ester | 10/S/ ester | 10/U/ ester | 10/E/ ester | 12/S/ ester | 12/U/ ester | 12/E/ ester | 8/S/ ether | 8/U/ ether | 8/E/ ether | 10/S/ ether | 10/U/ ether | 10/E/ ether | 12/S/ ether | 12/U/ ether | 12/E/ ether |
|---|---|---|---|---|---|---|---|---|---|---|---|---|---|---|---|---|---|---|---|---|---|---|---|---|---|---|---|
| 8/S/sorb | 1.000 | | | | | | | | | | | | | | | | | | | | | | | | | | |
| 8/U/sorb | 0.080 | 1.000 | | | | | | | | | | | | | | | | | | | | | | | | | |
| 8/E/sorb | 0.003 | 0.189 | 1.000 | | | | | | | | | | | | | | | | | | | | | | | | |
| 10/S/sorb | 0.187 | 0.658 | 0.083 | 1.000 | | | | | | | | | | | | | | | | | | | | | | | |
| 10/U/sorb | 0.008 | 0.325 | 0.738 | 0.156 | 1.000 | | | | | | | | | | | | | | | | | | | | | | |
| 10/E/sorb | 0.453 | 0.309 | 0.022 | 0.563 | 0.048 | 1.000 | | | | | | | | | | | | | | | | | | | | | |
| 12/S/sorb | 0.003 | 0.189 | 0.999 | 0.081 | 0.737 | 0.022 | 1.000 | | | | | | | | | | | | | | | | | | | | |
| 12/U/sorb | 0.130 | 0.002 | 0.000 | 0.006 | 0.000 | 0.026 | 0.000 | 1.000 | | | | | | | | | | | | | | | | | | | |
| 12/E/sorb | 0.385 | 0.369 | 0.030 | 0.647 | 0.063 | 0.905 | 0.029 | 0.019 | 1.000 | | | | | | | | | | | | | | | | | | |
| 8/S/ester | 0.215 | 0.599 | 0.069 | 0.934 | 0.134 | 0.621 | 0.068 | 0.007 | 0.707 | 1.000 | | | | | | | | | | | | | | | | | |
| 8/U/ester | 0.318 | 0.442 | 0.040 | 0.743 | 0.083 | 0.803 | 0.040 | 0.014 | 0.896 | 0.806 | 1.000 | | | | | | | | | | | | | | | | |
| 8/E/ester | 0.437 | 0.322 | 0.024 | 0.582 | 0.051 | 0.978 | 0.024 | 0.024 | 0.926 | 0.640 | 0.824 | 1.000 | | | | | | | | | | | | | | | |
| 10/S/ester | 0.405 | 0.350 | 0.027 | 0.620 | 0.058 | 0.934 | 0.027 | 0.021 | 0.970 | 0.680 | 0.867 | 0.956 | 1.000 | | | | | | | | | | | | | | |
| 10/U/ester | 0.972 | 0.075 | 0.003 | 0.176 | 0.007 | 0.433 | 0.003 | 0.139 | 0.366 | 0.203 | 0.302 | 0.417 | 0.386 | 1.000 | | | | | | | | | | | | | |
| 10/E/ester | 0.387 | 0.367 | 0.029 | 0.644 | 0.062 | 0.907 | 0.029 | 0.019 | 0.997 | 0.705 | 0.894 | 0.929 | 0.973 | 0.368 | 1.000 | | | | | | | | | | | | |
| 12/S/ester | 0.243 | 0.680 | 0.113 | 0.987 | 0.198 | 0.616 | 0.114 | 0.013 | 0.694 | 0.953 | 0.781 | 0.634 | 0.669 | 0.231 | 0.691 | 1.000 | | | | | | | | | | | |
| 12/U/ester | 0.722 | 0.160 | 0.008 | 0.332 | 0.019 | 0.692 | 0.008 | 0.063 | 0.606 | 0.374 | 0.519 | 0.672 | 0.633 | 0.698 | 0.609 | 0.394 | 1.000 | | | | | | | | | | |
| 12/E/ester | 0.048 | 0.000 | 0.000 | 0.001 | 0.000 | 0.008 | 0.000 | 0.631 | 0.005 | 0.002 | 0.004 | 0.007 | 0.006 | 0.052 | 0.006 | 0.004 | 0.021 | 1.000 | | | | | | | | | |
| 8/S/ether | 0.759 | 0.031 | 0.001 | 0.088 | 0.002 | 0.269 | 0.001 | 0.188 | 0.219 | 0.105 | 0.172 | 0.257 | 0.234 | 0.787 | 0.220 | 0.139 | 0.493 | 0.070 | 1.000 | | | | | | | | |
| 8/U/ether | 0.381 | 0.372 | 0.030 | 0.651 | 0.064 | 0.899 | 0.030 | 0.019 | 0.995 | 0.712 | 0.902 | 0.921 | 0.965 | 0.363 | 0.992 | 0.698 | 0.602 | 0.005 | 0.216 | 1.000 | | | | | | | |
| 8/E/ether | 0.889 | 0.060 | 0.002 | 0.145 | 0.003 | 0.374 | 0.002 | 0.168 | 0.314 | 0.169 | 0.256 | 0.360 | 0.332 | 0.917 | 0.316 | 0.197 | 0.620 | 0.066 | 0.875 | 0.311 | 1.000 | | | | | | |
| 10/S/ether | 0.316 | 0.007 | 0.000 | 0.023 | 0.000 | 0.083 | 0.000 | 0.601 | 0.064 | 0.028 | 0.049 | 0.078 | 0.070 | 0.333 | 0.065 | 0.042 | 0.177 | 0.318 | 0.443 | 0.064 | 0.387 | 1.000 | | | | | |
| 10/U/ether | 0.685 | 0.033 | 0.001 | 0.087 | 0.002 | 0.250 | 0.001 | 0.264 | 0.205 | 0.103 | 0.163 | 0.239 | 0.218 | 0.711 | 0.206 | 0.128 | 0.448 | 0.113 | 0.899 | 0.203 | 0.790 | 0.548 | 1.000 | | | | |
| 10/E/ether | 0.389 | 0.365 | 0.029 | 0.641 | 0.062 | 0.911 | 0.029 | 0.020 | 0.994 | 0.702 | 0.890 | 0.933 | 0.977 | 0.371 | 0.996 | 0.688 | 0.612 | 0.006 | 0.222 | 0.988 | 0.318 | 0.066 | 0.208 | 1.000 | | | |
| 12/S/ether | 0.057 | 0.000 | 0.000 | 0.002 | 0.000 | 0.009 | 0.000 | 0.687 | 0.007 | 0.002 | 0.005 | 0.009 | 0.007 | 0.062 | 0.007 | 0.005 | 0.025 | 0.939 | 0.083 | 0.007 | 0.077 | 0.356 | 0.131 | 0.007 | 1.000 | | |
| 12/U/ether | 0.415 | 0.012 | 0.000 | 0.035 | 0.001 | 0.121 | 0.000 | 0.477 | 0.095 | 0.043 | 0.073 | 0.115 | 0.103 | 0.435 | 0.096 | 0.061 | 0.244 | 0.236 | 0.572 | 0.094 | 0.499 | 0.849 | 0.681 | 0.097 | 0.267 | 1.000 | |
| 12/E/ether | 0.204 | 0.622 | 0.073 | 0.959 | 0.142 | 0.598 | 0.073 | 0.007 | 0.684 | 0.974 | 0.781 | 0.617 | 0.656 | 0.192 | 0.681 | 0.977 | 0.357 | 0.002 | 0.098 | 0.688 | 0.160 | 0.025 | 0.096 | 0.678 | 0.003 | 0.040 | 1.000 |

Note: All p-values shown as 0.000 signify < 0.001.

Tables 11 and 12. Tukey's method identifies many more significant differences than does the LSD. For example, although the ANOVA determined that the three-factor interaction for dispersion effectiveness was highly significant ($p < 0.001$; see Table 6), the LSD test was unable to identify any differences between adjacent pairs of means when they were rank ordered, whereas Tukey's method identified 104 significant differences when all possible pairwise comparisons were evaluated (Table 12).

## CONCLUSION

In this research, dispersant effectiveness was measured using two methods: measurement of the droplet-size distribution using an optical particle counter (OPC) and extraction of the dispersed oil into dichloromethane followed by gravimetric measurement of the mass of extracted oil. The gravimetric method produced systematically lower estimates of dispersion effectiveness than the size-distribution method, probably due to evaporation of volatile oil components during evaporation of the solvent. Therefore, treatment effects were evaluated based on the size-distribution data.

All dispersants produced bimodal droplet-volume distributions, and the two major modes had similar mean diameters (about 4 μm and 10 μm). This bimodal size distribution may be due primarily to the characteristics of the baffled flask test, although experiments conducted in salt water using the same dispersants tend to produce trimodal volume distributions (unpublished data). In the latter case, an additional large volume mode (mean diameter of about 25 μm) is produced. Despite the similar characteristics of all of the dispersed-oil volume distributions, dispersant formulations tended to change the distribution of oil between the two droplet-size modes. The relative amount of oil in the large and small size modes was quantified using the diameter of mean volume (DMV). Although the overall differences among DMV for dispersant formulations was relatively small, the dispersant characteristics exerted highly significant effects on DMV with all three main effects, all three two-factor interactions (*i.e.,* HLB x dispersant composition; HLB x surfactant chemistry; dispersant composition x surfactant chemistry), and the three-factor interaction (*i.e.,* HLB x dispersant composition x surfactant chemistry) being significant at the 95% confidence level. The strong statistical significance (ANOVA results in Tables 6 and 7) of these treatment effects and interactions appears to be due to the high reproducibility of the droplet-size distributions that were produced by each dispersant formulation. Dispersion effectiveness was much more variable than were the size distributions. It is also of interest to note that divalent cations seem to exert a more important influence than monovalent cations in terms of droplet size distribution effects from the use of dispersants (Belk, *et al.*, 1989)

Dispersants formulated with sorbitan ester surfactants and with HLB of between about 8 and 10 exhibited the best performance when the dispersion of a weathered Mars Blend crude oil was tested in the synthetic lake freshwater. At least two of the dispersant formulations that were tested were highly effective with >90% of the added oil transferred to the aqueous phase in the baffled flask test. Dispersants formulated with fatty acyl ester or ether surfactants and higher HLB were relatively ineffective. The least effective dispersants transferred less than 60% of the added oil to the aqueous phase. It is unknown to what extent this observation applies to other types of freshwater formulations.

This report showed that, at least with Mars Blend crude oil in simulated lake water, dispersants can be designed to drive an oil slick into the freshwater column with the same efficiency as in saltwater as long as the HLB is optimum. Clearly, many more oils would need to be tested under different conditions (temperature, organic content, water composition, etc.) to enable firm conclusions that oil can be dispersed in freshwater as a response tool. The ultimate decision to use dispersants in treating freshwater petroleum oil spills is up to the federal on-scene coordinator, the incident command team, the regional response teams, and EPA Headquarters, since many other factors need to be considered before rendering a decision to disperse oil into the water column. It is beyond the scope of this report to consider such factors. Its purpose was simply to determine if freshwater dispersion is possible and to determine whether effective freshwater dispersants can be designed to produce stable oil droplet distributions in such an environment.

## REFERENCES

Belk, J. L., Elliott, D. J. & Flaherty, L. M. (1989). The comparative effectiveness of dispersants in fresh and low salinity waters. In: *Proceedings, 1989 Oil Spill Conference*, 333-336. American Petroleum Institute. Washington, DC.

Blondina, G. J., Singer, M. M., Lee, I., Ouano, M. T., Hodgins, M., Tjeerdema, R. S. & Sowby, M. L. (1999). Influence of salinity on petroleum accommodation by dispersants. *Spill Science and Technology Bulletin, 5(2)*, 127-134.

Bruheim, P., Bredholt, H. & Eimhjellen, K. (1999). Effects of surfactant mixtures, including Corexit 9527, on bacterial oxidation of acetate and alkanes in crude oil. *Applied and Environmental Microbiology, 65(4)*, 1658-1661.

Bruheim, P. & Eimhjellen, K. (2000). Effects of non-ionic surfactants on the uptake and hydrolysis of fluoresceindiacetate by alkane-oxidizing bacteria. *Canadian Journal of Microbiology, 46(4)*, 3 87-390.

Clayton, J. R., Payne, J. R. & Farlow, J. S. (1993). *Oil Spill Dispersants: Mechanisms of Action and Laboratory Tests*. CRC Press, Inc. Boca Raton, FL

Clark, J. (2004). *Dispersant basics: Mechanism, chemistry, and physics of dispersants in oil-spill response.* Presented to the National Research Council Committee on Oil-Spill Dispersants: Efficacy and Effects, March 15.

Fingas, M., Bier, I., Bobra, M. & Callaghan, S. (1991). Studies on the physical and chemical behavior of oil and dispersant mixtures. In: *Proceedings, 1991 Oil Spill Conference*. American Petroleum Institute. Washinton, DC.

Fingas, M. F., Kolokowski, B. & Tennyson, E. J. (1990). *Study of oil spill dispersants effectiveness and physical studies.* Thirteenth Arctic and Marine Oilspill Program Technical Seminar, Edmonton, Alberta, Canada, Environment Canada.

Georges-Ares, A. & Clark, J.R. (2000). Aquatic toxicity of two Corexit dispersants. *Chemosphere, 40*, 897-906.

Lehtinen, C. M. & Vesala, A. M. (1984). Effectiveness of oil spill dispersants at low salinities and low water temperatures. In: *Oil Spill Chemical Dispersants: Research, Experience, and Recommendations*, 108-121. T. E. Allen. American Society for Testing and Materials. Philadelphia, PA.

Myers, D. (2006). *Surfactant Science and Technology*, Third Edition. John Wiley and Sons, Inc. Hoboken, NJ.

NRC (2005). *Oil Spill Dispersants: Efficacy and Effects*. National Academies Press. Washington, DC.

NRC, N. R. C. (1989). *Using Oil Spill Dispersants on the Sea*. National Academy Press. Washington, DC.

Porter, M. R. (1991). *Handbook of Surfactants*. Blackie and Sons, Ltd. Glasgow, Scotland.

Sorial, G. A., Venosa, A. D., Koran, K. M., Holder, E. & King, D. W. (2004). Oil spill dispersant effectiveness protocol. II: Performance of revised protocol. *Journal of Environmental Engineering (ASCE), 130(10)*, 1085-1093.

Stoffyn-Egli, P. & Lee, K. (2002). Formation and characterization of oil-mineral aggregates. *Spill Science & Technology Bulletin, 8(1)*, 31-44.

Van Hamme, J. D. & Ward, O. P. (1999). Influence of chemical surfactants on the biodegradation of crude oil by a mixed bacterial culture. *Canadian Journal of Microbiology, 45,* 130-137.

# CHAPTER SOURCES

The following chapters have been previously published:

Chapter 1 – This is an edited reformatted and augmented version of a National Commission on the BP Deepwater Horizon Oil Spill and Offshore Drilling publication, report *The Use of Surface and Subsea Dispersants During the BP Deepwater Horizon Oil Spill.*

Chapter 2 – This is an edited reformatted and augmented version of a United States Environmental Protection Agency, Office of Research and Development publication, report *Analysis of Eight Oil Spill Dispersants Using In Vitro Tests for Endocrine and Other Biological Activity,* dated June 30, 2010.

Chapter 3 – This is an edited reformatted and augmented version of a United States Environmental Protection Agency publication, report *Comparative Toxicity of Loisiana Sweet Crude Oil and Chemically Dispersed LSC to Two Gulf of Mexico Aquatic Test Species*, dated August 31, 2010.

Chapter 4 - This is an edited reformatted and augmented version of a United States Environmental Protection Agency publication, report *Comparitive Toxicity of Right Oil Dispersant Products on Two Gulf Of Mexico Aquatic Test Species*, dated June 30, 2010.

Chapter 5 – This is an edited reformatted and augmented version of a PDF downloaded from *www.epa.gov/bpspill/dispersants/qanda-phase2.pdf*, dated August 2, 2010.

Chapter 6 – This is an edited reformatted and augmented version of a Environmental Protection Agency publication, report R-08/037, dated March 2008.

# INDEX

## D



## E

## F



## G



## H



## I



## J



## L

## M



## N



## O



## P

## Q



## R



## S



## T

## U



## V



## W



## Y